SACRAMENTO PUBLIC LIBRARY
D0098554

Life's Ratchet

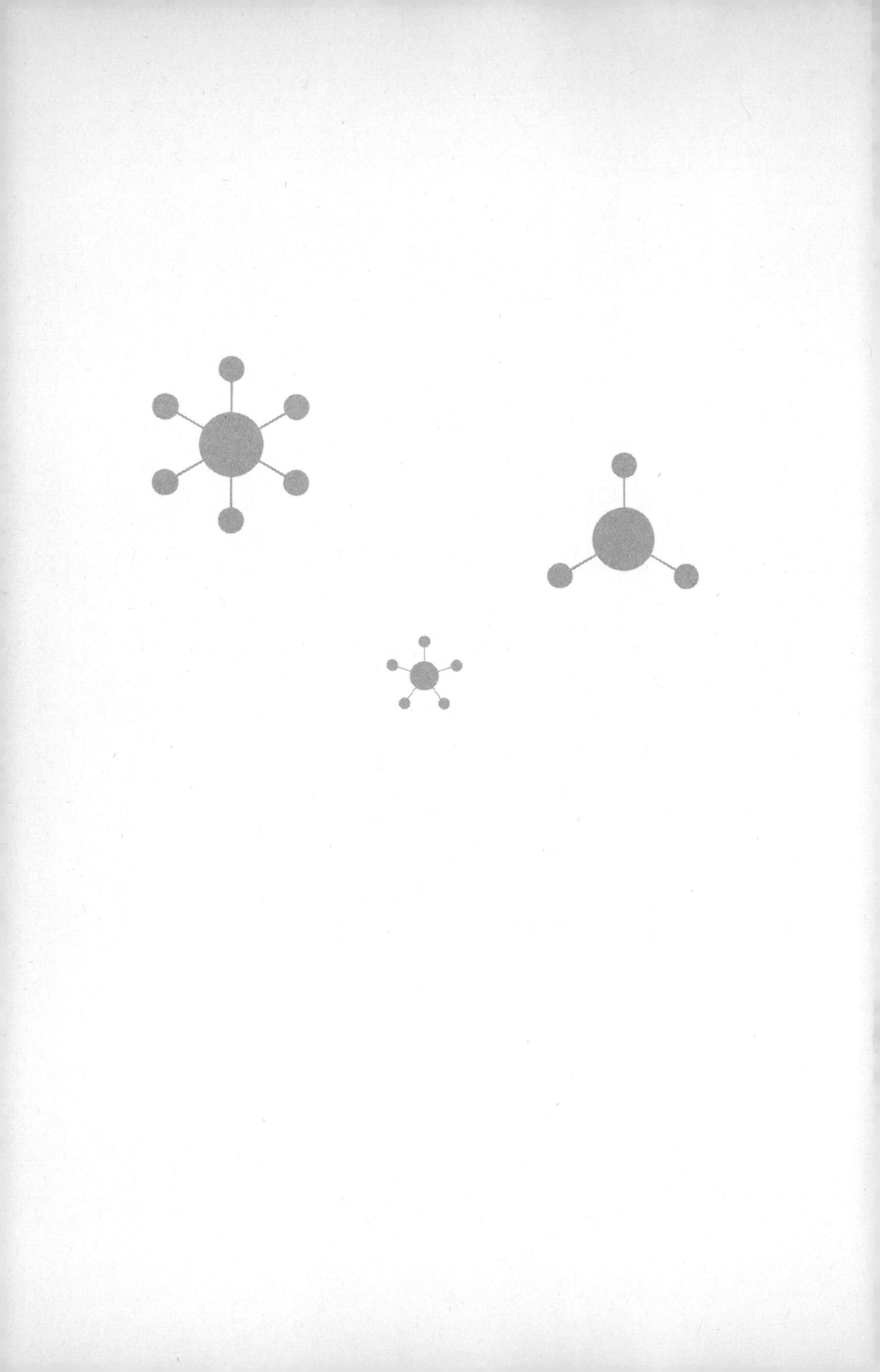

Life's Ratchet

HOW MOLECULAR MACHINES EXTRACT ORDER *from* CHAOS

PETER M. HOFFMANN

BASIC BOOKS
A Member of the Perseus Books Group
New York

Published by Basic Books,
A Member of the Perseus Books Group

Books published by Basic Books are available at special discounts for bulk purchases in the United States by corporations, institutions, and other organizations. For more information, please contact the Special Markets Department at the Perseus Books Group, 2300 Chestnut Street, Suite 200, Philadelphia, PA 19103, or call (800) 810-4145, ext. 5000, or e-mail special.markets@perseusbooks.com.

Designed by Pauline Brown
Typeset in 11.5 point Dante MT Std by the Perseus Books Group

Library of Congress Cataloging-in-Publication Data

Hoffmann, Peter M.
Life's ratchet : how molecular machines extract order from chaos / Peter M. Hoffmann.
p. cm.
Includes bibliographical references and index.
ISBN 978-0-465-02253-3 (hardcover : alk. paper)
ISBN 978-0-465-03336-2 (e-book)
1. Molecular biology. 2. Bioenergetics. 3. Life (Biology) I. Title.
QH506.H636 2012
572'.33—dc23

2012018626

10 9 8 7 6 5 4 3 2 1

To my lovely wife, Patricia,
and my parents, who raised me
to always want to know more.

Contents

Introduction

What Is Life?

Little Fly,
Thy summer's play
My thoughtless hand
Has brushed away.
Am not I
A fly like thee?
Or art not thou
A man like me?
For I dance
And drink, and sing,
Till some blind hand
Shall brush my wing . . .

—William Blake

That crude matter should have originally formed itself according to mechanical laws, that life should have sprung from the nature of what is lifeless, that matter should have been able to dispose itself into the form of a self-maintaining purposiveness—that [is] contradictory to reason.

—Immanuel Kant

A BLACK SPECK WHIZZES IN FRONT OF MY EYES. Absentmindedly, I swat at it, only to find I have killed a fruit fly. I probably should not be upset, but I have destroyed a living being, an autonomous, moving thing that makes its own decisions, flies around, finds its own food, and knows how to make more copies of itself. I have destroyed a marvelous machine created far beyond the capabilities of our best scientists and engineers. Now, as I look at the dead creature, I wonder: What made this motionless mass of water and organic molecules so happily alive just a moment ago? What does it mean when we say something is "alive"?

I am a physicist, not a biologist. To be honest, my formal biology studies ended when I was in eleventh grade, and I never took a single university-level class in the subject. Why write this book? When I was in high school, I loved science and mathematics, but I could never get too excited about biology. It seemed like a lot of tedious memorization and ad hoc theories and appeared to lack the coherence, clarity, and universality of physics. This remained my opinion for many years while I finished my undergraduate studies in Germany and took off for graduate school in the United States. For a while I was your typical arrogant physicist, getting a good chuckle out of Ernest Rutherford's quote: "Physics is the only science; all else is stamp collecting."

My conversion started when I was a doctoral student in materials science at Johns Hopkins University. My Ph.D. advisor, Peter Searson, was fascinated with a new, powerful instrument invented just eight years earlier: the atomic force microscope (AFM). Since he was not familiar with the AFM's operation, he put me and my friend Arun Natarajan in charge of figuring out how it worked. An AFM is a thousand times more powerful than the best optical microscope. Unlike conventional microscopes, AFMs do not use light to obtain images but rather visualize samples by touch: A tiny, sharp tip is moved across a sample, and the minute forces pushing on the tip are used to create an image. Tips are very sharp, only a few nanometers across, which allows for very small objects to be imaged.

One day, a fellow student brought in samples to image. He had deposited DNA molecules on a flat substrate and was wondering if our AFM could make them visible. We were blown away when we saw little worm-like strands appear on the computer screen—each a single DNA molecule, only two nanometers in diameter. We had touched the molecule of life.

Life took a few more turns before I finally converted to the wonders of biology. After a stint as a research fellow at Oxford University, I arrived at Wayne State University in Detroit as a fresh-faced assistant professor. Initially, I concentrated on what I knew: using AFM to look at atoms and molecules on surfaces and measuring forces between them. Two subjects fascinated me: the preponderance of randomness at the scale of atoms and the connections between the microscopic world of atoms and our macroscopic world. At the tiny scale of atoms and molecules, chaos reigns, yet at the scale of humans, order prevails (at least for the most part). How does this order arise? This is the subject of statistical mechanics, and in my research, I probed the transition from "noise" to "order" (and thus the limits of statistical mechanics) as I measured forces in small clusters of atoms and molecules.

As it happens often in life, new opportunities arise quite by chance. Another AFM researcher, Heinrich Hoerber, joined Wayne State University. Hoerber, a pioneer in new nanotechnology techniques applied to molecular biology, had been a postdoc with Gerd Binnig, the Nobel Prize–winning co-inventor of the AFM. I was fascinated with Hoerber's work, and when he subsequently left to take a position at University of Bristol in the United Kingdom, I inherited his Wayne State collaboration: to measure the motions of particular molecular machines implicated in the spread of cancer. Here was an opportunity to combine my interest in statistical mechanics and the tools of nanotechnology with something new: molecular biology. At the same time, I had the opportunity to contribute to a cure for cancer. So I beefed up on biology and started a new research direction. As I learned more about molecular biology, I discovered the fascinating science of molecular machines. I realized that life is the result of noise and chaos, filtered through the structures of highly sophisticated molecular machines that have evolved over billions of years. I realized, then, there can be no more fascinating goal than to understand how these machines work—how they turn chaos into life. This is the story I will share in this book.

What *is* life? Scientists have tried to answer this question for as long as science has existed. For Aristotle, the body was matter, but a soul was needed to give the body life. Even today, such views are common in the general public. Books like *The Secret* tell us we have vast untapped reserves of "life energy" that can help us attract riches and happiness. Yet, a special

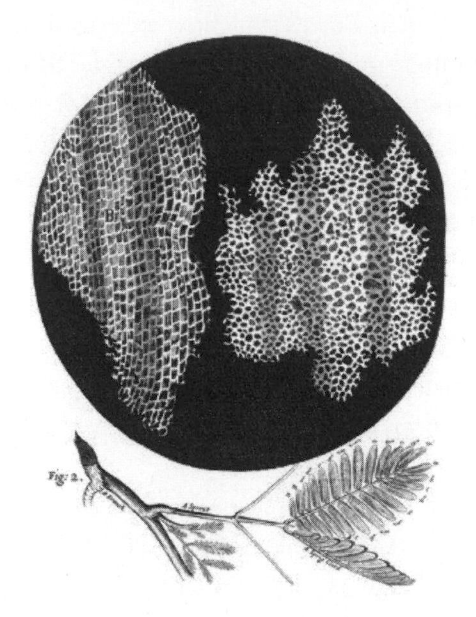

FIGURE 0.1. The first living cells observed and recorded. From Robert Hooke's *Micrographia*, 1665.

life force has never been detected. If we look at the balance sheet of energy intake (food) and output (motion, heat) of any living organism, there is no missing energy or untapped energy source.

At the other extreme we have the view of living creatures as complicated and intricate machines. The French philosopher René Descartes believed animals (but not humans) to be machines without souls. According to his view, animals did not feel pain. To explore animals' internal machinery, he promoted vivisection, a practice we find barbaric today.

Beginning in the seventeenth century, with the invention of the microscope, scientists searched for the secret of life at ever smaller scales. Biological cells were first described in Robert Hooke's *Micrographia* in 1665 (Figure 0.1). It took until 1902 for chromosomes to be identified as carriers of inheritance. The structure of DNA was deciphered in 1953, and the first atomic-scale protein structure was obtained in 1959. Yet, even while scientists dissected life into smaller and smaller pieces, the mystery of life remained elusive.

Is this reductionist approach doomed to fail? Many people, including many eminent biologists, think so. But I believe they are wrong. To be sure, reductionism is not enough: Many unexpected and important phenomena emerge only from the complex interaction of many parts. These emergent phenomena cannot be explained by looking at the parts alone. *Holism* (the understanding that the whole is more than its parts) is part and parcel of any explanation of life.

Nevertheless, the reductionist approach of looking at smaller and smaller pieces of living organisms has been a story of continued success. And it may finally be claiming a very big prize—one of the great mysteries of life: What creates "purposeful motion" in living beings? This was one of the original mysteries of life, formulated by Aristotle more than two thousand years ago. Aristotle assigned this motion to purpose. But today, having penetrated into the realm of molecules, we do not find purpose. Instead, we find random motion. Today, this great question has morphed into another question: How can *molecules* create the "purposeful" action that characterizes cells and bacteria? How do we go from assemblies of mere atoms to the organized complex motions in a cell?

In this book, we will find answers to these questions that have plagued science and philosophy for thousands of years. What kept us so long from solving this mystery? What we lacked were the right tools and concepts to study life at sufficiently small scales. How small is small enough? The secret of life's activity is found at the scale of a nanometer—a billionth of a meter.

Thanks to the advances of nanotechnology, we can now see the smallest parts of life at work: autonomously moving molecules performing specific tasks like tiny robots. Our cells are cities full of molecular-sized worker bees, who, like magic, built themselves, go where they are needed, do what they need to do and are recycled again. How can mere molecules move in specific ways to perform specific tasks? Are these amazing molecular robots imbued with a special life force? Are they controlled by a higher consciousness? Astoundingly, the force that drives life at the smallest scale is not a mysterious, supernatural force, but it is a surprising one nevertheless. The force that drives life is chaos.

As a newcomer to molecular biology and with the unique perspective of a physicist, I feel well suited to tell the story of the new discoveries of life at the nanoscale. I have not been in the field long enough to take anything for granted—everything is new and exciting, and I want to share this excitement with my readers. Yet, I owe this story to the wonderful researchers who have come before me: The biologists who painstakingly figured out the detailed pathways of cellular activity, the biochemists who identified the chemical nature of the molecular machines of our cells, and most recently, the physicists who are trying to find the general principles behind the hustle and bustle of our cells. The fundamental goal of this book is to follow the discoveries of these scientists and to find out what it takes to turn a molecule into a machine; and many molecular machines into a living cell.

When we follow the path of reductionism to understand life, the starting point of our quest must be the molecular scale. Deep down, life is a complex dance of molecules which can be understood in the context of physics. In 1945, the Nobel Prize–winning physicist Erwin Schrödinger predicted that genetic information, the blueprint to make a human, is coded in the structure of molecules. In his book *What Is Life?* he envisioned the genetic code to be contained in chemical "letters" as part of an aperiodic crystal (today we call it a polymer), and the size of each letter in the genetic code to be a few nanometers in size. These physics-based predictions inspired a young Francis Crick to decipher the mystery of DNA just thirteen years later. Crick and his coworker James Watson found Schrödinger's predictions to be quite accurate. Everything we have learned about life at the molecular scale has conformed to known physical principles. In this book, I follow the path Schrödinger first walked on and look at life from the point of view of a physicist.

Yet, even at the molecular scale, life is incredibly complex; without this complexity, life could not function. In 1970, another Nobel winner, the French biochemist Jacques Monod, concluded, in *Chance and Necessity*, that the complex machinery of our cells must be the result of an unbelievably lucky cosmic accident: "The universe was not pregnant with life nor the biosphere with man. Our number came up in the Monte Carlo game. Is it any wonder if, like the person who has just made a million at the casino, we feel strange and a little unreal?" Many scientists have embraced Monod's support for chance over necessity. They are concerned about opening the scientific floodgates to vitalism (the idea that life requires special forces) and religion. Necessity implies there is an external reason for life to exist. If there is such a reason, there must be a driving force outside physics or biology.

Other scientists saw things differently. In 1917, D'Arcy Wentworth Thompson, a British biologist and mathematician, published a unique book, *On Growth and Form*. Thompson showed how shapes of living plants and animals have analogues in the nonliving physical world. He argued that the shapes of our bodies are not due to chance, but are the *necessary* result of physical forces and geometrical constraint. Thompson found a way to favor necessity over chance without implying religion or vitalism. For him, the structure of the living organism was the necessary result of mathematics and physics.

As we enter the microscopic world of life's molecules, we find that chaos, randomness, chance, and noise are our allies. Without the shaking and rattling of the atoms, life's molecules would be frozen in place, unable to move. Yet, if there were only chaos, there would be no direction, no purpose, to all of this shaking. To make the molecular storm a useful force for life, it needs to be harnessed and tamed by physical laws and sophisticated structures—it must be tamed by molecular machines.

The fruitful interaction of chance and necessity also explains how these chaos-harvesting machines were "designed" by evolution. Chance and necessity may even explain how our minds work, how we have new insights, and why we have intuition. This book is a vindication for randomness, a much maligned force. Without randomness, there would be no universe, no life, no humans, and no thought.

Where does chaos come from? Why are atoms in perpetual random motion? The random motions of the atoms in our bodies are an afterglow of the creation of the universe, the big bang. The big bang created a universe full of energy, and, eventually, it created stars like our sun. With the sun as intermediary, the energy of the big bang shakes the atoms of our cells—making life on Earth possible.

Like it or not (and I hope you will like the idea by the time you have read this book), chaos *is* the life force. Tempered by physical law, which adds a dash of necessity, chance becomes the creative force, the mover and shaker of our universe. All the beauty we see around us, from galaxies to sunflowers, is the result of this creative collaboration between chaos and necessity. The potential for life was already written into the book of our universe as soon as physical law met the violent motions of elementary particles. For me, this insight makes the story of life a beautiful, even spiritual story.

Understanding life is not an easy task. The fundamental nature of life is one of the most enduring hard questions of science. Scientific literature is replete with articles that attempt to explain various aspects of life—yet much is still conjecture; much controversial. The public rarely hears about the exciting developments in science, because understanding requires advanced knowledge of biology, chemistry, and physics. To make matters worse, scientific literature is written in a language that makes it difficult even for scientists to understand each other. In this book, I will cut through the fog of scientific hieroglyphics and make the latest theories of life accessible to the intelligent reader. I do not have all the answers, and some things I write in this book will turn out to be wrong. But science is not an old, dusty book of settled facts. It is a living, breathing story of discovery, a true adventure of the human mind.

Let the adventure begin.

1
The Life Force

The human body is a machine that winds itself,
a living picture of perpetual motion.

—Julien Offray de La Mettrie

Come, said my soul, Such verses for
my Body let us write, (for we are one).

—Walt Whitman, *Leaves of Grass*

LIFE IS THE DANCE OF A BEE AND THE ROAR OF A LION. IT IS the tangle of a rain forest and the mortal battle between bacteria and host. Life is amoeba and elephant, evolution and extinction, and the power to transform a planet. The complexity and variety of life is staggering, but for physicists, life begins at a more basic level. All life started as a circle dance of molecules billions of years ago. The lion and the bee, the humble yeast and the mighty blue whale all share the same jittering molecules in their cells; we are all cousins.

But while life is based on molecules and energy, it seems to defy a purely physical explanation. When we look at a living being, we immediately recognize it as alive, as fundamentally different from a rock or a cloud. Yet, when we try to define life, we run into difficulties. There seems to be something indefinable, some special ingredient that separates inanimate matter from living flesh. When a loved one dies, we despair at not being able to recreate life. It is as though a special ingredient, a "life force," has left the body. Life seems forever beyond our powers and understanding.

And yet, we know that modern science has the power to manipulate life. From genetic engineering to brain imaging, science has penetrated living matter to its very core. The dichotomy between our everyday experience of the purposefulness and magic of life, and the fact that when we go looking for the magic ingredient, we only find matter and mechanism, has occupied human minds for thousands of years. It has led to a drawn-out battle between those who see purpose and those who see mechanism. In this battle, sometimes one side gained the upper hand, sometimes the other. Has the battle finally been decided? And if yes, who won?

The Secret

If I had to vote for the most abused scientific terms, *energy*, *power*, and *force* would be on the short list. According to the motivational speaker Bob Proctor, human beings are an incredible source of power and could use the "power in their body" to illuminate "a whole city for nearly a week."*

Since Proctor is so precise ("nearly a week"—why not a whole week?), it may be worth double-checking his calculations. It turns out that it is quite easy to calculate the power rating of a human being. Power, in physics, measures how *fast* energy is transformed from one form to another, and not the amount of available energy. Proctor is confusing power with energy. But I don't want to quibble about that. Let's pretend he means that the power rating of one human is equivalent to the hundreds of thousands of light bulbs that illuminate a city.

Humans transform energy from food into motion, heat, and thought. Energy is conserved. The energy we expend during a day comes from the food we eat. A typical energy intake from food is 2,500 food calories per day. One food calorie is equal to 4,184 joules of energy. A human consuming 2,500 food calories takes in approximately 10.5 million joules (2,500 calories × 4,148 joules) in energy from food a day. This sounds like a lot.

---

* Bob Proctor, motivational speaker, in *The Secret*, directed by Drew Heriot (Prime Time Productions, 2006). Quote from IMDb, "Memorable Quotes for *The Secret* (2006)," www.imdb.com/title/tt0846789/quotes, accessed July 8, 2010.

However, a day has 86,400 seconds, and therefore the *rate* at which our bodies transform this energy is 10.5 million joules divided by 86,400 seconds, or about 120 watts (where 1 watt = 1 joule per second). Far from illuminating a whole city, a human being has about the same power rating as *one* light bulb.

Humans talk, write, walk, and love using the same amount of energy per second as a light bulb, a device that does nothing but shine light and get hot. This amazing fact, far from denigrating humans, is a testament to how efficient a human body is. But even more importantly, it is a testament to the wondrous complexity of our bodies, which can do so much with so little.

Humans and other living beings are not sources of energy. We are consumers of energy, taking high-grade energy in the form of food and releasing it in the form of low-grade heat into the environment. When we stop eating, we starve and die. This simple truth is nothing new, yet books like Rhonda Byrne's *The Secret* (which claims that "human beings manage their own magnetizing energy") sell millions of copies, making us believe that there are untapped sources of energy within us.* Why is this idea so persuasive? Where did this notion of life force or energy come from?

The idea that life is infused with special energies or forces is as ancient as humanity itself. When people today are attracted to books like *The Secret*, it may be because the idea of a life force is deeply engrained in our psyche. For at least a hundred thousand years, humans have tried to bridge the gulf between life and death by placing flowers, food, or tools in burials with their departed. For our ancestors, death was an unnatural state, as all of nature seemed to be ever changing, moving, and alive.

Nature's powers of motion and change were associated with *anima*, the soul. Animism, the belief that all of nature was alive and governed by

* Rhonda Byrne, *The Secret* (New York: Atria Books, 2006). Certainly, everything is energy. Byrnes's next sentence, after the claim "human beings manage their own magnetizing energy," goes even more astray: "You are an energy magnet, so you electrically energize everything to you." The statement make little sense and would not be found in any physics book. But making scientific sense matters little when it comes to publishing books. After all, *The Secret* sold millions of copies.

spiritual forces, survived the centuries and was part of respectable European philosophizing well into the twentieth century.* The use of "magic" crystals and magnets for healing is still part of some people's beliefs today, as they believe that these items have special energies that affect life and health. In animism, not only animals, but also rocks, the wind, the river, were alive. In such a belief system, the concept of dying did not make much sense. The ancients believed that when a person died, he or she was not really dead, but instead the person's spirit had moved somewhere else. It was important to supply the dead person with tools and gifts for this new existence. Our now familiar distinction between living beings and lifeless matter evolved much later. Once this happened, most things—rocks, water, air—were recognized as lifeless, and *living* became a mystery in need of explanation. Living matter was now seen as being substantially different from all other matter and had to be endowed with extraordinary forces or a soul. We call such a belief *vitalism*.

Vitalism began with the Greeks, most notably with the philosopher and scientist Aristotle (384–322 BCE). For Aristotle, life was different from inanimate matter because it had "soul"; it was "animated." It is not a coincidence that we identify the word *animated* with being in motion. Purposeful motion, which includes locomotion, growth, and internal motion of the organism, was for Aristotle (and still is today) the most conspicuous attribute of life.

Aristotle spent a lot of time thinking about the soul, as recorded in his book *De Anima*. He identified several problems with defining soul: Is the soul a whole, or is it made of parts? Are there different types of souls for horses, dogs, and people? What is the soul's relationship to the body? Aristotle realized there was a problem distinguishing the soul from the body: "Are all affectations of the complex of body and soul, or is there any among them peculiar to the soul itself? . . . there seems to be no case in which the soul can act or be acted upon without involving the body." But

* As I trace the history of beliefs and knowledge about life, I will restrict myself to Western philosophy and science. This is not to diminish the thoughts and achievements of other cultures, but Western thought provides many of the examples of how to think about life. And despite the many important outside influences on Western thought, modern science is mostly an outgrowth of European philosophical history, starting with the ancient Greeks.

FIGURE 1.1. Democritus—typically shown as the bearded, laughing philosopher. In the history of science, he certainly had the last laugh. After all, he was (mostly) right about atoms.

despite raising this issue, Aristotle never addressed the question about the necessity of the soul. In fact, he would have found such a question absurd—for him and his fellow Greeks, the existence of a soul was self-evident: "Knowledge of the soul . . . contributes greatly to the advance of truth in general, and, above all, to our understanding of Nature, for the soul is . . . the principle of . . . life."

Atomism

When I learned about Greek philosophy in high school, I first noticed all the things the Greeks got wrong. To my teenage self, it seemed naive to think of earth or fire as elements. In truth, the Greeks made enormous progress, from their belief in Olympian gods Zeus and Hera to their model of nature using four elements. The Greeks were also the first people to base ideas on scientific observations. The philosopher Anaximenes of Miletus (585–529 BCE) determined that air was the fundamental element, because it could be rarefied or condensed. He based this idea on his observations of evaporation, condensation, drying, and wetting. Other observations that guided Greek philosophy included the growth of seeds (Anaximander), breathing (Anaximenes), fossils in rock (Xenophanes), the necessity of water for life and buoyancy (Thales), and the random motion of suspended dust (Democritus; Figure 1.1).

Aristotle was, without a doubt, the most prolific (and most scientific) of all Greek philosophers. In *De Anima*, he provided a comprehensive overview of what his predecessors thought of the mystery of the soul. Based on meager experimental evidence, ancient philosophers ventured surprisingly close to modern ideas: According to Aristotle, the philosopher Democritus imagined the soul as a fire consisting of myriads of jostling particles, which Democritus called "atoms." Democritus got the idea for the incessant motion of atoms from observing the random movements of dust grains in beams of sunlight. As we will see, the ceaseless motion of atoms and molecules plays a central role in our modern understanding of life—a motion we can rightly call a *molecular storm*.

Unfortunately, Aristotle found the ideas of Democritus and Pythagoras absurd. For him, movement is the result of thought and will, not the random motion of atoms: "Democritus says that . . . atoms . . . owing to their ceaseless movements draw the . . . body after them and so produce its movements. . . . we may object that it is not in this way that the soul appears to originate motion in animals—it is through intention or process of thinking."

Thus, from the Greek philosophers to the early-nineteenth-century biologists, there were three possible solutions to the problem of explaining life: Assume an overarching, universal principle that determines the purpose of the entire universe (animism); assume a special life force that distinguishes life from matter, thus reserving purpose for life alone (vitalism); or deny purpose altogether (mechanism, atomism). All of these approaches had their problems. Animism erased the clear distinction of the inanimate and the alive, vitalism gratuitously introduced an unseen force and raised the additional question of how this force interacted with the body, and atomism seemed impotent to account for those of life's activities that seemed to show clear purposefulness, such as growth and reproduction.

Atomism made a brief resurgence with the philosopher Epicurus (341–270 BCE) and was later revived by the Roman philosopher Lucretius (99–55 BCE) in his famous poem *De rerum natura*. Explaining the universe as the result of atomic motion, Epicurus invented the "swerve"—the sudden, random swerving of atoms that otherwise would move on straight, predictable paths. The swerve explained how atoms clumped together or bounced off each other. It explained creation, spontaneity, chance, and free will. While the idea of the swerve seems gratuitous, Epicurus understood that an atomistic explanation of the universe needed a mixture of necessity and chance.

However, the difficulty of reconciling the random motion of atoms with the obvious purposefulness of life doomed atomism for many centuries and has cast a long shadow on our understanding of life until today. The battle of soul versus the atom continued to be central to understanding life, even though the terms changed and biological knowledge became more refined.

For the most part, the ancients vacillated between vitalism and animism. Aristotle clearly understood that life was special and did not postulate souls for rocks and mountains. He did, however, think of motion as due to a purpose. A rock "wanted" to fall down, because it was made of the element earth and wanted to go back to the earth. On the other hand, the Stoics (a school of philosophy founded by Zeno of Citium circa 300 BCE) believed in a more animistic world, where a mysterious ordering force, the *pneuma* ("breath") gave rise to all existence. The pneuma was like an ancient version of "the force" in *Star Wars*, the fictional energy field created by all living things.

With the rise of Christianity, both the atomism of Epicurus and the animism of the Stoics became discredited. For the early Christians, Plato's philosophy, which was based on the transcendent world of ideas and not our material reality, was much more palatable. Plato's universe was the result of reason, not chance. Steeped in Platonic philosophy, the evangelist John wrote: "In the beginning was the Word, and the Word was with God, and the Word was God." Following John, the early Church, and especially St. Augustine, equated this "Word" or reason with God, and the world of ideas with heaven. The material world was relegated to a corrupted reflection of the spiritual world, which contained the real truth, the truth of God.

Many writings of Aristotle, lost in the West for nearly a thousand years, were saved by the Muslims. In the twelfth century, his works reentered Western philosophy. The early schools of theology in Paris, Oxford, Toledo, and Cologne, which later became universities, were stunned when they encountered the comprehensive knowledge contained in Aristotle's numerous writings, from logic to physics, statecraft to biology. Unlike Plato, Aristotle saw the material world as primary, and ideas as mere generalizations of observed objects and phenomena. In this, he was quite close to what scientists believe today.

While Aristotle's ideas threatened the established neoplatonic theology of the time, he could not be ignored. His philosophy was too comprehensive and too well reasoned to be dismissed. A new philosophy, *scholasticism*, was born to reconcile Aristotle's philosophy and science with Christian theology. Not everything in Aristotle's books was counter to Christian beliefs. For example, he clearly dismissed atomism and the accompanying idea of chance as an important player in the universe. For Aristotle, the most important force was purpose. Motion also required an explanation and could not be attributed to the unexplained random motion of atoms. Instead, Aristotle postulated a first mover—which St. Thomas Aquinas, the most famous of the twelfth-century scholastic philosophers, equated with God. In living beings, the soul was the prime mover. According to Aristotle, "[the soul] acts and [the body] is acted upon, and the [body] is moved and the [soul] moves."

Aristotle's concept of the soul has survived until today, and is evident in the catechism of the Catholic Church: "'Soul' also refers to the innermost aspect of man, that which is of greatest value in him, that by which he is most especially in God's image: 'soul' signifies the spiritual principle in man. . . . The unity of soul and body is so profound that one has to consider the soul to be the 'form' of the body: i.e., it is because of its spiritual soul that the body made of matter becomes a living, human body; spirit and matter, in man, are not two natures united, but rather their union forms a single nature." This idea of the soul as the form of the body, which renders it alive, comes from Aristotle. The Catholic catechism contains two meanings to the word *soul*: Aristotle's life-giving "form" and a "spiritual principle." Even though these two meanings are often conflated, they are radically different concepts.

Modern physics, chemistry, and evolution can explain what makes a cell or an organism alive and what gives it "form." Bodies are complex assemblies of interacting cells, operating according to an evolved program written in the cell's DNA. A soul is not needed as the source of form, locomotion, nutrition, or reproduction (in contrast to what Aristotle thought). The concept of soul may make sense in the second meaning—as a noncorporeal, unique essence of a human being; as a shorthand to aspects of a living being that encompasses personality, dignity, intelligence, mind, and the connection to others.

The Christian adoption of Aristotelian ideas, while at first quite radical, put a straightjacket on science for many centuries. The Greek philosophers welcomed debate, and it is reasonable to assume that Aristotle would have been horrified to learn that his musings were now taken as gospel.

Medicine and Magic

The vitalistic ideas of the Greek philosophers, as well as their early penchant for scientific observation, profoundly influenced the practical science of life: medicine. Originally based on magic and faith-healing, medicine was put on a more rational footing by Hippocrates and other Hippocratic thinkers around the time of Aristotle. Medicine became a rational science, based on an understanding of the universe. Greek (and, later, Roman) medicine culminated in the ideas of Galen (129–217 AD), whose books dominated Western medicine for fourteen hundred years. As late as 1559, a member of the London College of Physicians had to publicly rescind his comments when he dared to criticize Galen in front of his colleagues.

Galen's medicine, based on Aristotle's philosophy and Stoic ideas, was heavily vitalistic: It was loosely based on the four Greek elements, which he called the body's vital fluids (or "humors"). Galen believed that the pneuma, the "life spirit" that circulated in the air, entered the body through the lungs. In the heart, the pneuma mixed with blood (one of the four humors) and produced the "vital spirits," which were responsible for movement. As part of the soul, these vital spirits were associated with heat, which, according to Galen, was generated in the heart when blood mixed with air. The connections between air, heat, and soul were a recurring theme from ancient Greece to the dawn of the scientific age. And as we will see, thinking of heat as the "living power" is not so far from reality as you might think.

Ancient medicine combined observations (in Galen's case, mostly by dissecting animals) with a philosophical understanding of the universe. For the ancient physicians, life was associated with heat, and heat was generated by fire (one of the four traditional elements), which must be nourished by air (another of the elements). This kind of reasoning sometimes came close to the right answers, but ultimately, the ancients were victims of their own philosophical predilections. Without the methods of modern science—controlled experiments, the testing of hypotheses and quantitative arguments—medicine remained in a rut for more than fourteen hundred years.

In the Renaissance, when the study of human nature took center stage, medicine, astronomy, and physics finally broke out of the cage of Aristotelian and Galenic thought. The Renaissance was a time of rediscovery and reassessment, during which scholars combed the globe for ancient manuscripts. It was not the discovery of modern scientific methods that allowed Renaissance physicians to break with Galen and Aristotle; it was the discovery of ancient *magical* manuscripts. The arguments of the ancient magicians were based on the correspondence between the human body and the universe as a whole, and they led to the development of alchemical medicine. For example, in his book *Of Natural and Supernatural Things*, the Benedictine monk and alchemist Basilius Vesalius provides a recipe for "Spirit of Mercury . . . which cures all diseases, be it dropsie, consumption, gout, stone, falling sickness, apoplexy, leprosy, or howsoever called in general"—a recipe, surely, that would be more likely to cause apoplexy than to cure it.

The main proponent of this new medicine was the Swiss physician Paracelsus (1493–1541), born Phillippus Aureolus Theophrastus Bombastus von Hohenheim. (While apparently the word *bombastic* is not based on Hohenheim's middle name, he certainly was that: "Let me tell you this: every little hair on my neck knows more than you and all your scribes, and my shoe buckles are more learned than your Galen and Avicenna, and my beard has more experience than all your high colleges.") Even though Paracelsus's alchemical medicine was often no better than the old Greek medicine, his decisive break with the ancient Greek tradition and his emphasis on using chemistry were an important step forward. At times, Paracelsus sounded amazingly modern: "Medicine is not only a science; it is also an art. It does not consist of compounding pills and plasters; it deals with the very processes of life, which must be understood before they may be guided."

The various approaches to medicine—traditional and herbal medicines, Galenic medicine based on the balance of humors, and the new alchemical medicine of Paracelsus—coexisted and were vigorously debated, often on what appears to us today as dubious grounds. Yet, that they *were* debated ushered in the era of modern critical science.

The Mechanical Philosophy

The battle between ancient and alchemical medicine was raging when a new idea emerged: the idea that human bodies were merely machines and that the body's function could be understood as the workings of discrete parts. An inadvertent hero of the mechanical view of life was the English physician William Harvey (1578–1657), a thoroughgoing vitalist who still believed in Galen's vital spirits. Yet Harvey was the first to understand the true function of the heart. According to Galenic medicine, the heart was the source of heat in the blood and the place where blood mixed with air to create "vital spirits." Galen believed that the arteries originated in the heart, and the veins in the liver. The arteries and the veins were separate systems, connected only through the porous septum in the heart.

Although little evidence supported the porous nature of the septum, few physicians were brave enough to criticize Galen's theories. The famous physician Andreas Vesalius (1514–1564), who published some of the most influential and detailed books on human anatomy, could not find any porosity in the heart; nevertheless, in the first edition of his *De fabrica* he accommodated Galen: "The septum is formed from the very densest substance of the heart. It abounds on both sides with pits. Of these none, as the senses can perceive, penetrate from the right to the left ventricle. We wonder at the art of the Creator which causes blood to pass . . . through invisible pores." But by the second edition of his book, Vesalius had to admit, with some regret, that there was simply no way that Galen's theory could be correct.

The path was now clear for Harvey, who, like Vesalius before him, had studied at the University of Padua. The way Harvey disproved Galen was enormously influential: Going beyond dissections and observations, Harvey used a simple *quantitative* argument, which was unprecedented and powerful. If the heart was the source of blood, the total amount of blood generated in the heart could easily be estimated by multiplying the volume of the heart with the rate of pumping. This would result in 540 pounds of blood every hour—a giant amount. Where would it all go? The only reasonable explanation was that a limited amount of blood circulated through the body; and whatever the heart pumped out, came back to the heart a short time later. Harvey's mathematical reasoning had an enormous impact on the subsequent history of the life sciences: Life, like the rest of nature, could yield to quantitative analysis and, with it, careful experimentation.

Despite Harvey's modern scientific methods, his work drew praise from the Paracelsian physician Robert Fludd (1574–1637). Fludd saw in the circulation of blood a confirmation of his alchemist views that the macroscopic world of the stars was reflected in the microscopic world of the human body: As the planets go around the sun, so the blood circulates around the heart. But Harvey's findings also received nods from more modern scientists and philosophers, especially René Descartes (1596–1650), who in *The Description of the Human Body* vigorously championed Harvey and argued that the body was a machine.

Descartes's philosophical ideas were an important step toward a mechanical view of life. Performing (sometimes gruesome) experiments on animals, he discerned that the body acted like a machine with pumps and pipes. He was one of the first natural philosophers to argue for the investigation of the body from a mechanical perspective, devoid of any mysterious forces. These ideas landed Descartes in hot water with the Catholic Church, despite his being a devout churchgoer. Descartes tried his best to reconcile what he saw in nature with Catholic theology. His solution was to divorce "mind" from the mechanical worldview he espoused. The mind or spirit was to be the realm of the soul and the divine, while the body was pure machine. The soul, which once explained everything, from the growth of plants to the human mind, had now been confined to mind alone. Everything else was matter in motion.

The first modern atomists, after the almost complete suppression of atomism during the Middle Ages, were Isaac Beeckman (1588–1637), a Dutch philosopher and scientist, and Pierre Gassendi (1592–1655), a French Jesuit priest. Beeckman, Descartes's teacher and friend, was considered one of most educated men of Europe at the time. Gassendi followed Beeckman in arguing for a revival of atomism and strived to make Epicurean atomism palatable for the Catholic Church. Gassendi had a different approach from that of Descartes: Whereas Descartes created *dualism*, the separate realms of matter and soul, Gassendi animated his atoms with the power of God, returning to an animistic view of the universe. If atoms were responsible for life, the necessary intelligence had to be built into them. And who endowed the atoms with this intelligence? God, of course.

This idea was later adopted by the German philosopher Gottfried Leibniz (1646–1716), who replaced atoms by "monads," atomic units of thought. Later materialist philosophers, however, would discard God altogether and instead endow atoms with uncreated purpose, an idea that during the Enlightenment, the French philosopher Voltaire (1694–1778) found laughable. According to Voltaire, the idea of some kind of uncreated intelligence inherent in atoms was ridiculous. Wasn't it simpler to just believe in God? After two thousand years of debate, it seemed that philosophers had not advanced much beyond Aristotle and Democritus.

But this would be unfair: The mechanical worldview, the revival of atomism, and the combination of rational examination and experiment were the foundation for one of the most influential periods in the history of science, the scientific revolution, which lasted from the late sixteenth to the eighteenth century. During this period, modern science was born and natural philosophers began to distinguish science from mysticism. The towering figures of this period were Galileo Galilei (1564–1642) and Isaac Newton (1642–1726). Both Galileo and Newton were atomists and believed in using experiment and reason to find new truths about nature. Newton had a Lucretian idea of how atoms form macroscopic matter; he thought of matter as made of small, hard particles that stick together to make larger particles. This idea explained how materials can break apart, as they break "not in the midst of solid particles, but where those particles are laid together, and only touch in a few points."

Newton stood at the threshold between Renaissance mysticism and modern science, and much has been made of his interest in alchemy and obscure theological pursuits. Yet his alchemy led him to value experimental approaches and validated his atomism. Newton appeared to know the difference between mysticism and science and kept his alchemy and theology neatly separated from his scientific and mathematical writings. He even went so far as to defend his scientific findings from those who thought that he was advocating new occult forces: "These Principles I consider, not as occult Qualities, supposed to result from the specifick Forms of Things, but as general Laws of Nature, by which the Things themselves are form'd; their Truth appearing to us by Phænomena, though their Causes be not yet discover'd."

The mechanical philosophy and the new atomism compelled scientists during the scientific revolution to look ever more closely at the living world, and advances in optics provided new instruments for the search of the "atoms" of nature. The microscope was invented in the late 1500s in the Netherlands by two Dutch spectacle makers, Zacharias Janssen and his son Hans. Improvements to the microscope were completed by Galileo (1609) and Cornelius Drebbel (1619). In 1614, Galileo observed that flies had "fur." Others observed mites and studied the structure of a fly's eye. The most famous early book on microscopic observations was Hooke's *Micrographia* of 1665. Robert Hooke (1635–1703), a master experimenter, used his homebuilt microscope to look at everything from flees to "gravel" in urine. The early microscopists discovered what Hooke called "small machines of nature," from the legs of flees to single-celled animals. Hooke was the first to see cells in cork. The first animal cells, red blood cells, were discovered shortly thereafter by Antonie Philips van Leeuwenhoek (1632–1723), but neither he nor Hooke realized that these cells were the smallest units of all living beings.

Hooke and his contemporaries discovered that life was a wondrous menagerie of mechanisms, from the smallest "animalcules" to the body of a human. The search for smaller and smaller units continued for over two hundred years, leading to the cell theory in the mid-1800s. For Hooke and his fellow microscopists, the mechanical philosophy compelled them to look carefully at the components of living beings. What they saw confirmed their mechanical view of life. Observing the growth of mold, Hooke noted: "I must conclude, that as far as I have been able to look into the nature of this Primary kind of life and vegetation, I cannot find the least probable argument to perswade me there is any other concurrent cause then such as is purely Mechanical."

L'homme machine

Some people just don't know when to shut up.

Julien Offray de La Mettrie (1709–1751), Brittany native, medical doctor, and radical Enlightenment philosopher, certainly didn't (Figure 1.2).

FIGURE 1.2. Julien Offray de La Mettrie. Another laughing materialist philosopher, although unlike Democritus, he has no beard.

When his first venture into philosophy, *A Natural History of the Soul* (1745), was burned in Paris for its impiety, he followed with an attack on the less-than-competent physicians of France. Having thoroughly upset both the medical and the religious establishments, he fled to Holland. Then, in 1747, he continued his attacks with *The Vengeful Faculty*, against the physicians, and *Man a Machine* (*L'homme machine*), against the priests. Holland was no longer safe, and La Mettrie found refuge at the court of Frederick the Great in Berlin. The Prussian king loved the French and the Enlightenment, and you couldn't be more Enlightenment than La Mettrie.

What had La Mettrie written in *L'homme machine* that so upset his contemporaries?

After receiving his medical doctorate from the University of Rheims, La Mettrie had served as medical officer to the French Guards and participated in a number of bloody battles. Through this experience, he developed a profound distaste for the slaughter of war and saw what savagery and injury could do to the human mind. He came to realize that reason and emotion, supposedly part of the soul, could be thoroughly altered by injury. Didn't this clearly show that Descartes's last refuge of soul—the mind—could ultimately be explained by pure mechanism? No wonder La Mettrie's more religious contemporaries were displeased.

Because of his provocative writings, La Mettrie is sometimes seen as the bad boy of the Enlightenment. The rumor that he died while overeating expensive pheasant pâté did not help his reputation (although it seems likely he died of food poisoning). However, under the combative veneer of his writing, there was a thoughtful philosopher and one of the most uncompromising representatives of the mechanical philosophy.

La Mettrie was a better provocateur and philosopher than scientist. But we can hardly fault him for that: While his explanations of procreation or "irritability" (see next paragraph) seem to us naive or laughable, he based them on what little was known about human physiology at a time when alchemical and ancient Greek ideas were still common. But his philosophy did not depend on such details.

At the heart of his most famous work, *L'homme machine*, were two observations: First, the functions of body and mind could be greatly altered by physical influences and therefore could not be independent of them. Second, living tissue, such as muscle, could move on its own, even when removed from the body. Experiments performed during La Mettrie's time demonstrated that isolated tissues could move when "irritated." This so-called irritability indicated to La Mettrie that life possessed "inherent powers of purposive motion." He put irritability at the center of his arguments, providing a list of ten examples, some quite gruesome, such as the frantic fluttering of headless chickens. In light of these observations, La Mettrie concluded that we cannot divorce the functions of the body or the mind from their physical nature. Instead, the functions must be the *result* of the physical and mechanical makeup.

La Mettrie's denial of the soul led to his being charged as an atheist (the standard charge for all philosophers who spoke out against Church doctrine). But he was more of a sincere agnostic: "I [do not] question the existence of a supreme being; on the contrary, it seems to me that the greatest degree of probability is in its favor. But that doesn't prove that one religion must be right, against all the others; it is a theoretical truth that serves very little practical purpose." For him, metaphysical and theological speculations about the soul served little purpose when a meal could make the "soul" happy and content and when we could see "to what excesses cruel hunger can push us." La Mettrie, tongue-in-cheek, observed that "one could say at times that the soul is found in our stomach." Observing that hunger, injury, drugs, and sleep affected people's minds, he felt certain that the soul was just part of the body, even if he could not explain in detail how it worked: "It is folly to waste one's time trying to discover its mechanism. . . . There is no way of discovering how matter comes to move."

La Mettrie was wrong to state that "there is no way" to discover how organized matter moves, but we can agree with him that a supernatural soul is probably not needed to render us alive, as Aristotle believed, because the soul is so dependent on the physical state of the body. La Mettrie concluded that the "soul's abilities" clearly depend on the "specific organization of the brain and the whole body." Therefore, the soul was nothing other than the organization of the body and, as a separate concept, empty. It had taken two thousand years before somebody could freely acknowledge that a soul was not necessary to explain the motions of the body.

La Mettrie's agnosticism gave him a modern outlook on the methodology of science (although modern scientists are more diplomatic) and a somewhat religious awe of nature governed by necessity. He wondered "why would it be absurd to believe that there exist physical causes for everything that has been made." Was it not "our absolutely incurable ignorance of these causes that has made us resort to a God?" For him, the answers for the mystery of life and mind lay not in chance, nor in God, but in nature.

La Mettrie's rejection of chance was very much in the spirit of the pre-Darwinian world. It was difficult to see what role chance could play in the emergence of the organized state of matter we call life. La Mettrie believed that a purely physical explanation of life was possible—but a mechanical explanation had to explain all aspects of life previously explained by religion. When God became an "unnecessary hypothesis," nature had to fill the gap and produce life and the human race out of necessity. The mechanical philosophy, like other explanations that relied on the fashionable ideas of the day, compared living beings to mechanical contraptions such as clocks.* Clocks, when well maintained, leave nothing to chance; they are the epitome of necessity.

---

* Other examples are explanations based on steam engines (thermodynamics) and chemistry or computers (the brain as computer). All of these explanations capture some aspects of life at least metaphorically, but ultimately they fail and should be used only with a giant grain of salt. The heart is not a clock, and the brain is not a computer.

For La Mettrie, nature acted on many levels, from the simple to the complex. He opposed artificial barriers to physical explanations and rejected the idea that there was a wall beyond which physics could not tread. The difference between a falling rock and a human mind was not one of different matter or different laws of nature, but one of a tremendous increase in complexity. And even if this complexity prevented us from completely explaining how a human mind works, this failure did not warrant the insertion of soul or God into this gap in our understanding. "Just as, given certain physical laws, it would not be possible for the sea to not have its ebb and flow, the same . . . laws of motion . . . would form eyes which see, ears that hear, nerves that feel, tongues that can or cannot talk depending on their organization; and finally, would fabricate the organ for thought. Nature has made in the human machine another machine, which finds itself capable of retaining ideas and creating new ones . . . it has made, blindly, eyes that can see; it has made without thought, a machine that thinks."

Animal Heat

While the mechanical philosophy dealt with the problem of motion quite well, and the vitalists with growth and reproduction, the problem of heat generation by animals remained a tough nut to crack. Why was heat so important? Most living things are warm, and when they die they grow cold. Clearly, heat had something to do with life. Already Pythagoras and Democritus considered heat, or more precisely "innate heat," the key to understanding life. For the ancient Greeks, the problem of heat was (literally) at the heart of the mystery of life. They believed that a kind of fire existed in the heart and that the lungs were needed to cool the heart's fire and remove exhaust. Plato writes in his *Timaeus*: "In the interior of every animal the hottest part is that which is around the blood and veins; it is in a manner an internal fountain of fire, which we compare to the network of a creel, being woven all of fire and extended through the centre of the body, while the outer parts are composed of air." This thinking persisted for centuries—even Harvey still believed that the heart was a source of heat and that the blood circulation he had discovered was a way to distribute heat throughout the body.

Such ideas were based on reasonable inferences from observations: Clearly, the heart resided near the center of the body—which seemed like a good place to put a stove—and it distributed warm, life-giving blood throughout the body. As for how the heart generated this heat, it was understood that heat was generally associated with fire. Thus, it seemed that some kind of "slow fire" in the heart generated the heat of the body. The ancients also guessed correctly that life's heat must be fueled by food. "[Food] is used up by our heat as oil is by a flame," Galen observed.

Galen and the ancients knew that both fire and life are extinguished in the absence of air, but they did not know why. Galen lamented: "if we could discover why flames are in these cases [when deprived of air] extinguished, we should perhaps discover what advantage the heat in animals derives through respiration." This blueprint for further research was not taken seriously by natural philosophers until fifteen hundred years later. In the early sixteenth century, animistic ideas about the nature of innate heat were still rampant. The French physician Jean François Fernel (1497–1558) believed that the innate heat enters the body once the embryo becomes an individual. He seemed to confuse innate heat with the religious idea of a soul. Descartes can be credited with bringing heat back into the realm of science, although his enthusiasm for mechanistic explanations led him astray when he speculated about the role of heat in the heart. While he believed in Harvey's blood circulation, he also believed the heart's motion was caused by an expansion of the blood due to the intense heat in the heart, and not by muscular contractions.

The primacy of heat as the central principle of life was challenged by the Belgian physician Jan Baptist van Helmont (1579–1644), who pointed out that frogs are quite cold, but also quite alive. Moreover, animals did not die when heat left them, but the heat left when animals died. Van Helmont was fond of chemical explanations, in the tradition of Paracelsus, and therefore saw heat as the result of chemical processes in the body, not the cause of such processes. He broke with the ancients, who had believed in heat as being innate, that is, inherent to living beings, like a soul. Van Helmont saw heat as a phenomenon that could be explained, as long as the right causes were identified.

The first experiments designed to answer Galen's question were performed by Robert Boyle (1627–1691), John Mayow (1641–1679), and

Robert Hooke—English mechanical philosophers and fellows of the still young Royal Society of London. They established that something in the air was involved in both fire and respiration. In both cases, heat was generated. Thus, the air drawn through the lungs did not cool the heat in the heart, but rather was an ingredient to produce heat in the first place. The mechanical philosophers speculated about "nitrous spirits" in the air, which would combine with sulfurous compounds in the blood in some kind of fermentation, generating the heat. Their science was still mingled with alchemy, and more sophisticated experiments were needed to solve the mystery.

The lack of a sound chemical explanation for innate heat allowed the pendulum to swing back to mechanical explanations in the eighteenth century. La Mettrie's teacher, the Dutch physician Herman Boerhaave (1668–1738), believed that the heat of the body was generated by friction when the blood was forced through the arteries. Many other physicians and scientists, thoroughly steeped in the mechanical philosophy of the time, shared this idea. However, these mechanical views were heavily criticized by the vitalists. An Edinburgh physician, John Stevenson, wrote in his 1747 paper "Cause of Animal Heat": "Not content with the ingenious . . . application of levers, ropes and pulleys to the bones, muscles and tendons; . . . millstones were brought into the stomach, flint and steel into the blood vessels, hammer and vice into the lungs. But all to no good purpose; there being certain bounds beyond which mechanical principles and demonstrations do not reach." The friction theory of animal heat was disproved by experiments, including an experiment by Benjamin Franklin in 1769. Franklin showed that fluids running through blood vessels could never generate enough heat. He resorted to chemical explanations and, observing that fermenting fruit had almost the same temperature as a living being, resurrected the fermentation theory of Hooke and Boyle.

Progress had to await a better understanding of the origin and nature of heat, just as Galen had predicted fifteen hundred years earlier. Because air seemed to be central to heat *and* respiration, natural philosophers tried to understand the chemical effect of air on various processes, from life to the oxidation of metals (or "calcination," as they called it, not knowing of oxygen). Confusingly, heat and fire sometimes released different kinds of "air" and changed the nature of the substance being heated or burned. Was this air "fixed" inside the substances and released upon burning? How did the air get to be trapped inside a solid? As long as air was still considered an element, it was difficult to understand why different airs disappeared in some reactions, while others emerged.

The mystery of fire seemed solved when the German (al)chemists Johann Joachim Becher (1635–1682) and Georg Ernst Stahl (1659–1734) developed the phlogiston theory of fire. According to this theory, phlogiston was a substance that was released upon burning, respiration, and calcinations, and this release generated heat. This theory was widely accepted until well into the eighteenth century. However, problems with this theory quickly emerged: For example, when a metal was heated in air, it turned into its "calx" (oxide). Phlogiston theory predicted that the calx would be lighter than the metal, because the calx formed when the phlogiston was released from the metal. But experiments showed the opposite to be the case: The calx was heavier than the metal. The proponents of the phlogiston theory explained this curious fact by claiming the phlogiston had levity, that is, some kind of antigravity. Such desperate explanations invited ridicule from the anti-phlogiston faction. The leader of this faction was the French chemist Antoine Lavoisier (1743–1794), who together with his wife and fellow chemist, Marie-Ann Pierrette Paulze (1758–1836), vanquished the phlogiston theory.

Lavoisier was a French nobleman who made a living collecting taxes (a job that landed him on the guillotine in 1794) and who, "on the side," revolutionized chemistry. It is little exaggeration to call Lavoisier the Newton of chemistry. The first person to clearly recognize the different chemical natures of the various airs, he discovered that water, earth, and air were compounds or mixtures. These discoveries thus eliminated most of the elements of the ancients—elements that were still taken seriously in the eighteenth century. The only element remaining after Lavoisier's overhaul of chemistry was fire or heat (which he renamed "caloric"). The demise of this last element was not far behind: At the dawn of the nineteenth century, Benjamin Thompson showed that heat is a form of energy (more about this in Chapter 3).

Lavoisier's most important experiment on animal heat consisted of measuring the heat output of a guinea pig versus that of burning coal, in relation to the carbon dioxide ("acide crayeux") generated during respiration or combustion. This was a tricky experiment: To measure the heat output, he used an ice calorimeter, invented jointly by him and Pierre-Simon Laplace (1749–1827). The instrument consisted of an isolated bucket with hollow walls stuffed with ice. Heat was measured by the amount of ice melted by whatever process was taking place inside the box. To avoid melting by the surrounding air, these experiments were done in the winter, when the temperature was just barely above freezing. The output of carbon dioxide was measured by placing the coal or the animal into a bell jar, so the air could be collected and analyzed. Lavoisier found that breathing and combustion generated roughly the same amount of heat for the same amount of carbon dioxide released. Respiration was now proven to be "slow fire." Not knowing about cells, Lavoisier placed this fire in the lungs.

While Lavoisier did not discover where and how heat was generated, he ushered in the age of quantitative physiology and established the connection between biology and thermodynamics. These findings, and his development of a rational chemistry based on true chemical elements, their compounds and mixtures, earns him a place among the greatest scientists of all time. When the French revolutionaries condemned him to death in 1794, the presiding judge declared: "The Republic needs neither scientists nor chemists; the course of justice cannot be delayed." The mathematician Joseph Louis Lagrange saw it differently: "It took them only an instant to cut off his head, but France may not produce another such head in a century."

This gruesome end to the life of a scientific genius can serve as a warning to all who see no value in supporting science. With Lavoisier's death, Germany and England quickly overtook France in the sciences (and in industry).

Mechanism Fails

Dissatisfaction with scholasticism, with blind adherence to Greek philosophy, and with mysticism of all kinds gave rise to the mechanical philosophy. As champions of rational thought and experimental evidence, the mechanical philosophers revolutionized science and philosophy. In physics, chemistry, and astronomy, they marched from triumph to triumph.

By the late eighteenth century, however, the limitations of the mechanical approach were starting to show. The mechanical picture of life, while informed by observations and experiments, turned out to be bloodless and impotent. The functions of the body, the interactions of organisms, the development of life, and the life of the mind did not yield to purely mechanical analogies. The specter of purpose was difficult to exorcise. The "purpose" of an acorn was to make an oak; the "purpose" of a brain was to think. How did the acorn know that it should become a tree? How could a lump of gray matter think? Could a clockwork explain such mysteries?

This state of affairs led to a schizophrenic approach to life: On one hand, a mechanistic approach had undeniably yielded important insights into how life worked; on the other hand, this approach seemed insufficient to account for *what* made matter alive. The basic attribute of life, its self-sustaining, self-organizing activity, remained outside the grasp of purely mechanistic explanations. Out of this tension, modern biology arose.

Germany and other central European countries, as well as Russia, became the center of biological research in the nineteenth century. The word *biology* was coined during this time. The German biologists developed sophisticated methods to study the development of embryos and the function of organs and muscles. The Germans improved microscopy and sample preparation techniques and studied the functions of the sense organs. In these studies, they accepted a mechanical picture of life's processes to a point, but were very aware that living organisms were not at all like a clock or a steam engine. There had to be something else.

Embryology, in particular, was a problem for the mechanists. Studying the development of a chick or a tadpole in detail, biologists saw matter take form, and a complex living being emerge from a lump of cells in a mysterious unfolding. You could not explain such a miraculous process with levers, pumps, and randomly moving atoms. It was absurd. Clearly, organic structures served a purpose. Living organisms had to be fundamentally different from anything encountered in the inanimate world.

The vexing problem of embryology expressed itself in the debates between the *preformationists* and the *epigenesists*. These debates revealed some surprising fault lines between the mechanists and the vitalists. Preformation was the idea that every living being had to be preformed in the egg or sperm. If you were a preformationist, you had to believe that before little Annie was born, there already existed a tiny version of Annie in the germ cells of her parents, too small to be detected by a microscope. Once activated by her father's sperm, the tiny embryo started to grow in her mother's womb, thus there was no necessity of unformed matter to acquire form, as the form was already present in the ovum of the mother. Surprisingly, preformationism was often the favored position of the mechanists, who we may naively consider the more scientific of the two camps. Yet, preformation seemed nonsensical: Taken to its logical conclusion, all of humanity, all those billions of humans who ever lived, had to already exist in Eve's ovaries as tiny versions of themselves.

Epigenesis, by contrast, claimed that unformed matter was shaped into a complex living being during embryonic development. This also posed problems. What directed this development? How could a lump of undifferentiated matter turn into a chicken or a frog? Mechanism could not explain how such spontaneous self-organization would work, even in principle. Epigenesis was the favorite of the vitalists. The formation of biological complexity from unformed matter was a task equal only to their beloved vital force—indeed, it was proof that such a force had to exist.

The dissatisfaction with mechanism led to a more romantic philosophy of life. Proponents of *Naturphilosophie* ("nature philosophy") saw a vital force acting throughout nature, striving to bring about higher and higher forms of being—a notion that sounded very much like animism. However, many German biologists were averse to explanations that seemed to invoke supernatural forces. While they allowed that a vital force was necessary, this vital force needed to be a part of nature and subject to scientific inquiry.

Trying to steer a path between mechanism and vitalism, the German philosopher Immanuel Kant (1724–1804) and his friend, biologist Johann Friedrich Blumenbach (1752–1840), created a new approach to the study of life: teleomechanism.* In 1790, Kant wrote to Blumenbach: "Your recent unification of the two principles, namely the physico-mechanical and

* The term *teleomechanism* was introduced in Timothy Lenoir, *"The Strategy of Life: Teleology and Mechanics in Nineteenth-Century German Biology* (Chicago, University of Chicago Press, 1989; 1982).

the teleological, which everyone had otherwise thought to be incompatible, has a very close relation to the ideas that currently occupy me." In the Aristotelian tradition, Kant and Blumenbach thought that life was directed by purpose (Greek *teleo*). In particular, Blumenbach espoused the existence of a formational drive (*Bildungstrieb* in German), which would provide the "means [by] which [the organisms] receive a determinate shape originally, then maintain it, and when it is destroyed repair it where possible." Blumenbach, and the biologists who came after him, did not see this vital force as a separate entity from the organism. Rather, it was a force contained within an organism, a result of its special organization and structure.

This view of self-contained special forces in organically organized bodies helped shape biology into an autonomous science. With the renewed recognition that life was special, life scientists could develop their own methods, ideas, and approaches. The role of the biologist was to find the rules by which special vital forces acted on organic matter. The German biologist Carl Friedrich Kielmeyer (1765–1844) went as far as to create a set of Newtonian laws that the vital forces would obey. First, he identified the forces acting in living beings: sensibility, irritability, reproduction, secretion, and propulsion. Then, he came up with various laws for these forces (most of which were subsequently proven wrong) and called them "the physics of the animal realm."

The biological tradition founded by Kant and Blumenbach led to several important discoveries in embryology and physiology. One of the most profound discoveries was the cell theory—the recognition that all living beings were made of smallest units, called cells. This insight was long overdue—after all, Hooke had coined the term *cells*, for the little compartments he saw in cork, as early as the late 1600s. The cell theory is usually credited to Matthias Jakob Schleiden (1804–1881), a German botanist; Theodor Schwann (1810–1882), a German zoologist; and Rudolf Virchow (1821–1902), a German physician, although numerous other scientists, from several countries, contributed to the final theory.

The cell theory corresponded to a kind of biological atomism, with cells as the "atoms" of life. But cells were not indivisible like atoms. Microscopists discovered even smaller structures inside cells, starting with the cell nucleus, which was named by Robert Brown in 1833.

The age of the German vitalist biologists opened up new vistas for inquiry. The teleomechanists were superb experimentalists and founded new schools of embryology, developmental biology, botany, and physiology. If these scientists were misguided in their theoretical ideas about vital forces, we may forgive them: Their experimental results have generally withstood the test of time.

An Irritable Frankenstein

> With an anxiety that almost amounted to agony, I collected the instruments of life around me, that I might infuse a spark of being into the lifeless thing that lay at my feet. It was already one in the morning; the rain pattered dismally against the panes, and my candle was nearly burnt out, when, by the glimmer of the half-extinguished light, I saw the dull yellow eye of the creature open; it breathed hard, and a convulsive motion agitated its limbs.
>
> —MARY SHELLEY, *FRANKENSTEIN*

Irritability, one of Kielmeyer's five vital forces, spurred the imagination of scientists, philosophers, and writers. One of them was La Mettrie, who was most impressed by the motions of muscles separated from their animal hosts. La Mettrie made irritability the center of his argument that life is pure mechanism, a complicated clockwork. But irritability also supported vitalist ideas: If muscles could move on their own, did this not prove the presence of a vital force? And could such a vital force, if distilled to its essence, not be used to bring life to dead tissue? At the height of the Romantic period, in the early 1800s, such speculation inspired a young woman with literary aspirations, vacationing with her lover in a villa on Lake Geneva, to write a novel.

Mary Wollstonecraft Shelley (1797–1851) conceived the idea for her novel *Frankenstein; or, The Modern Prometheus*, when she and her husband Percy Bysshe Shelley were holed up in the Villa Diodati with several friends, including the poet Lord Byron. It was 1816, the dreary "year without a summer," when a giant eruption of the Indonesian volcano Tambora darkened the European skies. During late-night chats, she and her friends talked about the increasingly gruesome experiments of scientists who were applying high voltages to dead animals and humans to demonstrate irritability. When her friends challenged each other to write a story or poem to entertain themselves, Shelley conceived of a gothic horror story about a scientist creating life from dead flesh.

The study of irritability, which La Mettrie had used as evidence for the mechanical nature of life, reached fever pitch when scientists of the day discovered that the newly founded science of electricity could be used to study living organisms. The iconic experiment of this era was conducted by Luigi Galvani (1737–1798), who attached a charged Leyden jar (a capacitor to store large amounts of electrical charge) to severed frog legs and observed that they kicked and twitched when electricity passed through them. Such experiments suggested that electricity could be the mysterious vital force philosophers had sought for centuries. In 1803, Galvani's nephew, Giovanni Aldini, went as far as to experiment with human corpses, making their faces twitch, their eyes open, and their extremities lift up. Such horrible experiments were supposed to show that "animal electricity" was a vital force responsible for the motion of muscles as it traveled along the nervous system. Unfortunately, what these experiments really inspired (much helped by Shelley's novel) was the idea of the overreaching, mad scientist—a figure that has since dominated much of the public's imagination.

I had not read Shelley's book until I started writing this book (although I was familiar with several movie versions, including the classic with Boris Karloff as the monster) and was surprised to find that Shelley never mentioned exactly how Dr. Frankenstein vitalized his creation. Shelley's hero deliberately keeps the reader in the dark, supposedly to prevent a repeat of the tragedy about to unfold: "I see by your eagerness and the wonder and hope which your eyes express, my friend, that you expect to be informed of the secret with which I am acquainted," explains Frankenstein in the novel, "that cannot be; listen patiently until the end of my story, and you will easily perceive why I am reserved on that subject." The giant switches and the lightning storm seen in various movie versions are all inventions of Hollywood, but it is clear that the studies of irritability, which were in the news in the early 1800s, provided the inspiration for Shelley's iconic novel.

Ever since its discovery by the Greeks (*electron* means "amber" in Greek, and amber generates static electricity when rubbed), electricity was considered a mysterious force and a subtle fluid. Such a mysterious force had to have some connection to the great mystery of life. Even Newton had suggested that electricity was responsible for animal motion. In the second edition of *Philosophiae Naturalis Principia Mathematica*, he speculated that the "subtle spirit" of electricity, transmitted through the nerves, "stimulated sensations" and "moved limbs."

Who was the inspiration for Shelley's Dr. Frankenstein? With animal electricity as the scientific object du jour, there were many candidates for this rather ignoble honor. Aldini, who passed large currents through the limbs of recently hanged criminals in front of large London crowds—to horrific effect—was certainly one of them. Another was Johann Wilhelm Ritter (1776–1810), a German scientist who preferred to apply the large currents to himself instead. Applied to his eyes, they made him see red and blue flashes, depending on which electric pole he had connected to his eyeball. Ritter died at a young age of unknown causes—but repeatedly electrocuting oneself cannot be too healthy.

The Conservation of Force—Or How Vitalism Was Vanquished by a Frog Leg

Although the late eighteenth to the mid-nineteenth century had become the age of teleomechanists and vitalists, by the mid-nineteenth century, mechanism had again gained the upper hand. This return to mechanistic explanations was mainly the work of two men: the English naturalist Charles Darwin (1809–1882), who destroyed teleology, and the German physiologist and physicist Hermann von Helmholtz (1821–1894), who vanquished the vital force.

Helmholtz was one of the last truly universal scientists. He made significant contributions to medicine, biology, and physics, in areas as diverse as heat in animals, irritability, the vital force, thermodynamics, electrodynamics, the conservation of energy, turbulence in liquids, and the physiology of the senses. His insights were groundbreaking, and most have withstood the test of time. He also invented several new experimental apparatuses, including the ophthalmoscope, the special microscope eye doctors point at your eyes to check the retina. His broad knowledge allowed him to make novel connections between different sciences. He could look at a system as complex as living tissue and determine the one parameter that linked it to the inanimate, mechanical world. He decided that this parameter was energy.

Physicists of the more arrogant sort often think that the interactions between physics and biology are purely one-way: Physics may explain biology, but biology has no bearing on physics. To cure such a misguided view of science, one should consider how Helmholtz came to argue for the law of energy conservation (or the conservation of force, as he called it). Helmholtz, trained as a physician, started his scientific career working on physiological experiments. It was these biological experiments that convinced him of the law of energy conservation.

Energy conservation had been in the air for a while. Descartes, Newton, and Leibniz had all argued for some quantity to be conserved in interactions between material corpuscles, although they could not agree on the type of conserved quantity (Newton argued for momentum or quantity of motion, while Leibniz argued for kinetic energy or *vis viva* ["the living force"]). Others had shown that work and kinetic energy could be converted into one another, for example, during free fall. Heat had been shown to be a type of motion, and it was already known that kinetic energy could be converted to heat through friction. In 1845, James Joule (1818–1889) showed that a fixed amount of work would result in a fixed amount of heat (what he called the mechanical equivalent of heat): "When equal quantities of mechanical effect are produced by any means whatever from purely thermal sources, or lost in purely thermal effects, then equal quantities of heat are put out of existence or are generated."

Drawing on his own biological observations and diligent studies of mathematical physics, Helmholtz extended energy conservation to all types of energy, thus declaring energy conservation a fundamental law of the universe. He showed how the conservation of energy can be mathematically derived from simple assumptions. While the mathematical treatment was Helmholtz's achievement alone, the idea of a universal law of energy conservation had been formulated some years earlier by another German scientist, Julius Robert von Mayer (1814–1878). Just like Helmholtz, Mayer was a physician venturing into physics and was also inspired by biology to proclaim the universal law of energy conservation. Helmholtz was unaware of Mayer's 1841 paper when he published his own ideas six years later. In his paper, Mayer repudiated vital forces, as

Helmholtz would do a short time later, and stated that "the cause of the chemical tension produced in the plant . . . is physical force." This physical force, or energy, as we would say today, was the same as the energy that would be obtained if we were burning the plant. Furthermore, this energy had to come from somewhere. If we postulated a mysterious vital force—a force that would require no source—we would be "carried . . . into unbridled fantasy," and all further investigation would "be cut off." No, said Mayer, the real explanation had to be that energy and matter were only converted from one form to another, and "that creation of either one or the other never takes place." In other words, even in something as complicated as a plant or an animal, energy was only transformed, but never created or destroyed. This is the universal statement of energy conservation.

In his famous essay of 1847, "Über die Erhaltung der Kraft" ("About the conservation of force"), Helmholtz, then only twenty-six, followed much the same line of argument that Mayer had set forth. Helmholtz felt that postulating mysterious vital forces added nothing to the investigation of how life works. Moreover, the presence of vital forces that could generate mechanical force from nothing would make it possible to construct a perpetuum mobile, a machine that generates energy from nothing. It was widely accepted that this was impossible. Energy conservation had to be correct, and special vital forces could not exist. Helmholtz showed that the law of energy conservation could be mathematically proven. He only needed to make the assumption that matter was made of pointlike particles, interacting through forces depending only on the distance between the particles. This mathematical proof and the expansive view of the new law met with some resistance from his older colleagues. Soon, however, new experiments proved Helmholtz and his fellow energy conservers correct.

Helmholtz was a dedicated mechanist from the beginning. Although he had studied with the influential researcher and teacher Johannes Peter Müller (1801–1858), who was part of the teleomechanist faction, he despised the very idea of invoking a vital force to explain anything. He designed several experiments to prove that vital forces were unnecessary to explain irritability. Most of these experiments, ironically, were to be performed on frog legs, the favorite experimental object of Galvani. Galvani, of course, had presented his frog leg experiments as support for vitalism.

In the first set of experiments, Helmholtz set out to prove that motion in muscles is caused by chemical processes, that is, that animal motion is a physicochemical process and is not related to any mysterious vital force. To prove this, he irritated frog legs several hundred times by passing electrical currents through them, just as Galvani had done. He then made several chemical extracts of the irritated frog legs and compared the extracts with extracts from non-irritated frog legs. He found that if the muscles had been irritated, a water-based extract lost mass and an ethanol-based extract gained an equivalent amount of mass. Clearly, some chemical compound in the muscles had been changed from a water-soluble to an alcohol-soluble form through the action of the muscles. This proved that the motion of the muscles caused a chemical change in the muscles, and Helmholtz concluded that muscles were machines that converted chemical to mechanical energy.

To establish that this energy was purely chemical, he next compared the heat that can be released upon chemical breakdown of food, called the latent heat, with the latent heat of excreted substances in animals. This was Lavoisier's experiment. However, since the time of Lavoisier, more-refined experiments had improved on Lavoisier's guinea pig. Helmholtz reviewed these experiments and concluded that the difference in energy between food and excrement accounted well for the observed animal heat. He was able to correct an error introduced by Lavoisier and the famous German chemist Justus von Liebig (1803–1873): Liebig (like Lavoisier before him) believed that the energy expended by an animal was exactly the same as oxidizing (burning) the animal's food in the oxygen the animal breathes. But the French physicist Pierre Louis Dulong (1785–1838) and the Belgian physicist César-Mansuète Despretz (1798–1863) had shown in their careful repeats of Lavoisier's experiments that an animal generated about 10 percent more energy than could be accounted for by the oxidation from respiration alone. This left an opening to the vitalists, who could point to the missing 10 percent as the contribution of the vital force. Instead, Helmholtz showed, the missing 10 percent came from the oxygen already contained in food, especially in carbohydrates and sugars. If this additional oxygen was included, food energy perfectly matched animal heat plus energy of the excrements, and no vital force was needed.

One more experiment was needed to completely eliminate the need for the vital force: Helmholtz had to show unequivocally that the energy to move muscles was contained in the chemical energy of the muscles (which they had received from food) and did not come from someplace else. At the time, it was known that a loss of the nervous system led to a cooldown of the body. Therefore, some biologists believed that the nervous system provided a source of vital force or animal heat. Helmholtz devised an ingenious setup to eliminate this last refuge of the vital force. By irritating three selections of tissue—a frog leg with its attached spinal nerve, a frog leg without the nerve, and the nerve without the leg—he ventured to show that any temperature increase due to the motion of the leg originated in the leg and was not due to any vital energies transferred to the leg from the nerves. To measure the minute temperature increase of a moving frog leg, he constructed a very sensitive device, comprising a thermocouple (a kind of electrical thermometer, which converts temperature into voltages), a magnetic coil to magnify the resulting voltage, and a dial that displayed the temperature after calibration of the device. His setup was accurate enough to record temperature changes as small as one-thousandth of a degree. Moreover, this setup was a physical representation of the law of energy conservation: Chemical energy in the frog leg (unleashed by electrical irritation) was converted into mechanical energy (motion of the leg), then into heat, electrical energy (thermocouple), magnetic energy (coil), and, finally, mechanical motion of the dial.

Helmholtz found that the presence of the nerves made no difference and that a nerve alone did not heat up at all. As long as the muscles moved the same amount, they heated up by the same amount, regardless of the presence of a nerve. In Helmholtz's view, the idea of a vital force was now untenable.

It would be too easy to conclude that the physicists carried the day. Yes, Helmholtz and others had shown that anything happening in a body—all the hallmarks of being alive, from animal heat to irritability—had to occur within the energy budget prescribed by the physicochemical world. If there was such a thing as a vital force, it had to be a force without energy and thus without potency. After Helmholtz, vital forces quickly fell from favor and have not been resurrected since, at least not in serious science. Nevertheless, for all of his single-minded eradication of the vital force, Helmholtz and his fellow physicists could not explain how unformed matter could organize into a complex organism. He had shown that it must happen within the law of energy conservation, but this did not explain how living beings formed. His research merely removed vital forces from the list of possible explanations.

Thanks to Helmholtz and others, biology returned to mechanism by the end of the nineteenth century, but not to the primitive, naive mechanism of the seventeenth-century mechanical philosophers. It was now clear that all biological processes occurred within the framework of chemistry and physics. The two disciplines were key to explaining physiological processes, such as irritability or animal heat. But it was equally clear that biology was fundamentally different from physics and biology: Complexity and development remained to be explained. Purpose was not yet exorcised.

Darwin and Mendel: From Chance to Purpose

The essential debates surrounding the mysteries of life never really changed. New findings brought the controversy over purpose versus mechanism into sharper relief, but the dilemma of biology in the late nineteenth century was fundamentally the same as the ancient debates between Aristotle and Democritus. By the 1850s, nobody could deny that to explain life's processes, physical, chemical, and mechanical forces had to be invoked. Yet mechanics seemed woefully insufficient to explain the extraordinary complexity and purposefulness of life. Mechanical explanations had become more powerful, and with the work of Helmholtz and his contemporaries, it was now clear that all forces or energies active in life were also present in inanimate matter. But this could not explain purpose. The dilemma of the physicists was the same dilemma Democritus had faced. How can complexity emerge from chaos?

The person who rescued mechanism from the bugbear of purpose was Charles Robert Darwin. Together with Alfred Russell Wallace (1823–1913), Darwin developed the theory of evolution based on natural selection. The idea of evolution was not entirely new. Ever since they noticed how different life forms were related to each other and to extinct forms, scientists had wondered if species could change over time. But without a satisfactory mechanism, the idea of evolution went nowhere.

Meticulously argued and written in a surprisingly accessible style, Darwin's revolutionary *Origin of Species* presented a strong case for natural selection as the driving force of evolution. Despite its general acceptance in modern science, Darwin's theory was absolutely earth-shattering and counterintuitive when it was published in 1859. When his contemporaries looked at marvelously designed plants and animals, Darwin's claim that these forms could have emerged by a blind, step-by-step process seemed outrageous.

Yet the theory was persuasive: Darwin's faithful supporter, the biologist Thomas Henry Huxley (1825–1895), exclaimed upon reading the book: "How extremely stupid not to have thought of that!" Indeed, like many brilliant ideas, Darwin's is exceedingly simple: All species exist as a population of individuals, each one a little bit different from every other. How these variations arose was unknown to Darwin (today we have a pretty good idea), but that they did exist was obvious. Over time, the variations that led to more reproductive success would surely increase in the population. In other words, individuals with a variation that allowed them to mate more successfully would have more offspring, and soon there would be more individuals with this particular variation. As conditions change or as populations are cut off from other populations, different variations will be favored and new species can emerge. Because the process is extremely slow, there are few opportunities to observe the emergence of a new species in a human lifetime.

Darwin's theory received serious opposition on religious grounds. In a popular book, *Natural Theology*, published in 1802, the theologian William Paley (1743–1805) had compared the complex mechanisms within living beings to the workings of a watch (as the rather less religious La Mettrie had done as well). Clearly, a watch assumed a watchmaker. It did not design and make itself. How could we even suggest that the infinitely more complex designs of living beings had made themselves? Paley's book was very popular at the time and was one of Darwin's favorites during his undergraduate years. It was only very gradually, and through thousands of painstaking observations, that Darwin came to realize that Paley was wrong.

By removing the last vestiges of purpose, Darwin's theory of evolution made God unnecessary for explaining the natural world. This, of course, was not the first time God had been made unnecessary. Physics had already banned God from the heavens ("I have no need for that hypothesis" answered the physicist Pierre-Simon Laplace when Napoleon asked him about the role of God in the universe), and Helmholtz had banned vital forces. The last refuge of the supernatural seemed to be the wondrous multiplicity of life forms. Now even that became the result of blind forces.

While much has been made of the religious critics of Darwin's ideas, his theory was similarly met with initial reservations by his fellow scientists. Here, the reasons were different. Darwin's theory lacked crucial ingredients: How were variations transmitted to the next generation, and how were new variations generated? Certainly, Darwin made a strong case, based on the breeding of dogs, orchids, and fancy pigeons, that animals can be changed dramatically over relatively short time. But this seemed to require the guiding hand of a breeder. Where did variety in nature come from? How could this variety lead to the changes observed? How were variations inherited? Why was the offspring similar, but not identical, to the parents? Although Darwin hit on the crucial mechanism to explain evolution—variation and natural selection—there were many gaps.

Unbeknownst to Darwin, a Moravian monk, noting curious patterns among generations of pea plants in his garden, had already discovered some of the missing puzzle pieces. The research of this monk, Gregor Mendel (1822–1884), was published in a rather obscure journal, the *Proceedings of the Natural History Society of Brünn*, and remained virtually unknown until it was rediscovered in the early twentieth century. What Mendel had discovered in his research with his beloved pea plants were the laws of inheritance: the fact that traits are inherited whole, and that traits from each parent can be combined in various ways in the offspring. Before his work, it was not clear how inheritance worked, and *blending* inheritance could not be discounted. In blending inheritance, the traits of the offspring are a blend of the traits of the parents. Mendel found this to be wrong: If it were true, pea plants would soon assume some average color, the blend of the colors of parents and grandparents. This is not what happened. If, for example a red pea was crossed with a white pea, Mendel got pink peas in the next generation, but surprisingly, the "granddaughter" peas could again be pure white or red. The trait white had not been blended out, but was merely dormant (recessive) and reappeared in a subsequent generation. The observation that the inheritance of traits was not blending, but rather was conservative, was extremely important for Darwin's theory. Only traits that could be passed on whole to the next generation could spread through a population and explain the emergence of a new species. If traits were blending, any new traits would soon be blended back into mediocrity.

Darwin's work was well received by the antivitalists of the time. Helmholtz embraced Darwin's ideas—they provided "the possibility of an entirely new interpretation of organic purposiveness." Indeed, together with the life force, purpose could now be placed on the ash heap of scientific history. But one problem remained. According to Helmholtz, "the controversy [over Darwin's theory] now centers mainly on the scope that should be assigned to variations of species." The problem was that the parts of Darwin's theory that dealt with *necessity* were no problem. *If* a new, useful trait emerged, Darwin's theory neatly explained how it would spread through a population by natural selection. This was just fine with the physicists—nineteenth-century physics was based on regularity and iron-clad natural laws. But what laws could explain where new traits came from? A life force? God forbid! Randomness? Not much better, and unimaginable to a thoroughgoing mechanist like Helmholtz. Indeed, the role of randomness in science had not much improved since Aristotle had rejected it over two thousand years earlier. Everything had to follow laws. But what was the law of variations?

2
Chance and Necessity

Everything existing in the universe is the fruit of chance and necessity.

—DEMOCRITUS

We believe that God created the world according to his wisdom. It is not the product of any necessity whatever, nor of blind fate or chance. We believe that it proceeds from God's free will.

—*CATECHISM OF THE CATHOLIC CHURCH*, ARTICLE 295

FAINT WISPS OF HYDROGEN AND HELIUM ARE SWIRLING through the immensity of space. The cosmos is vast and empty. Suddenly, the visitors glimpse a tiny illuminated island: a galaxy in a sea of nothingness. Looking closer, they notice that the island is made of smaller points of light: little nuclear fireplaces called stars, sprinkled into the cold darkness of space. Around many of the stars, small sand grains and blobs of gas leisurely circle their star's illumination. On one of the sand-grain planets, heated by its star to a comfortable 293 degrees Kelvin, white water vapor clouds beautifully set off the deep blue of saltwater oceans and the yellow-brown of continents. The tiny two-legged creatures inhabiting this little world, having just begun to glimpse a few feeble answers to the endless mysteries surrounding them, believe themselves to be the center of the universe. Chuckling, the visitors keep their giant spaceship cloaked, and move on.

Meanwhile on Earth, completely ignorant of being observed by powerful beings from outer space, humans seem incapable of shaking their illusions of superiority. As far as they know, this is the only planet with life, and they are the only intelligent life among millions of species. Only a few humans sense that this doesn't seem right. One of them writes a comic strip *Calvin and Hobbes*, in which a little boy called Calvin observes: "The best proof that there is intelligent life in the universe is that they have not contacted us." Maybe it's because they would die laughing.

In our belief that we are the center of the universe, we have assumed much, just to be proven wrong time and again: No, the solar system does not revolve around Earth. No, the universe does not end beyond Pluto, or even beyond our Milky Way galaxy, but it is much bigger than we ever thought, full of stars in some places, but for the most part filled with staggering emptiness. No, there is no special life force—our bodies are part of nature, run by molecules. And no, we are not a separate creation from all the other animals, but are their close cousins—all, including ourselves, historical accidents of evolution. In short, we are *lucky* to be here.

This last insult to our pride, that we may be here—at least partly—by accident, by chance, may be the toughest nut of all. But what is so bad about chance?

Randomness

You know the story. Traveling to some faraway destination, say, Turkey or Singapore, you run into the neighbor of your cousin, who happens to know about the perfect job opening for you. Fate, right? Many see hidden meanings when the unexpected strikes. Some become superstitious, some believe in karma, others in the will of God. Few people would admit it happened by chance, that it was simply a coincidence. Why do so many people reject the influence of randomness on their lives? Why are we bothered by randomness? Randomness invokes chaos, lack of control. If randomness rules, all bets are off. But is that really true?

Until the end of the nineteenth century, everybody believed that randomness had no place in any explanation of the world. People disagreed, however, how the specter of randomness was to be exorcised. The mechanists believed in mechanical necessity: If we knew the locations and speeds of all the particles in the universe at some point in time, and had a powerful enough computer, we could predict every future event. The religionists instead believed in the unfathomable will of God. If you couldn't explain why something happened, the explanation was that God wanted it this way. And the philosophers believed in . . . just about everything, except randomness. The consensus was that if something happened by chance, it only seemed that way because of our ignorance of all the circumstances.

Yet, clearly, many events seemed out of reach of our predictions. One way to deal with unpredictability was to accept it as fate, the will of God, or human ignorance. Another way was to find ways to quantify our ignorance, to tame the unpredictable.

A Short History of Gambling

The first time anybody ever thought about quantifying unpredictability was not in the name of science, but in the name of making a quick buck (or whatever currency they had at the time). The goal was to know how much to wager in a game of chance.

Gambling is an ancient pastime: Roman soldiers guarding the body of Jesus on the cross gambled for his meager belongings. Today, gambling is as popular as ever. Las Vegas, an entire city located in a place where there shouldn't be a city, is dedicated to this sinful activity. The city hosts poker championships, and there are always roulette games, blackjack, slot machines, and numerous other means to help you lose your money.

Poker has become something of a fad in recent years, especially among physicists. Unfortunately, my physics credentials are not much help: I am a lousy poker player. The problem is that I am terrible at bluffing. All good poker players can hide their emotions and are excellent strategists. But why should physicists in general (myself excluded) be any good at this game? The reason is that physicists understand probabilities.

As I understand poker, there are two ways of winning: getting a better hand than the other players, or making the other players believe you have a better hand. What makes a poker hand better? A poker hand is better if it is rarer, that is, if getting such a hand happens on average less often than getting a less valued hand. How often, on average, certain hands appear in a set of five cards selected at random is described by the *probability* of the hand. The probability of getting a royal flush—an ordered sequence (or straight) of cards of the same suit, starting with an ace—is 1 in 649,740. The probability of getting a pair (two cards of the same value) is much higher: 1 in 2.36. What does that mean? It means that in a very large number of randomly dealt five-card hands, a pair will occur about once every 2.36 deals, while a royal flush can only be expected every 649,740 deals. Of course, this does not mean that if you play 650,000 times, you can be guaranteed to be dealt a royal flush. Rather, it means that if you play billions or trillions of times, the number of royal flushes you are dealt divided by the number of total games would approach 1 / 649,740. But nobody has ever played poker a billion times. So how can we know?

The idea that different outcomes in games of chance, such as poker or dice, have different likelihoods, seems to be as old as gambling itself. How else would you determine how much to wager? Yet, for a long time, the probability of different outcomes was based on experience or feelings, rather than on quantifiable science.

In the movie *21*, Kevin Spacey portrays a morally impaired MIT mathematics professor, who teaches his students how to break the bank playing blackjack. Using card counting and secret signs, the students descend on Las Vegas casinos and clean them out. Certainly an interesting way to pay for college and a great idea for a movie (based on a true story), but it is hardly original: the idea to pay for college by gambling precedes *21* by five hundred years.

The person who first conceived of this ingenious use of gambling was born in 1501, an unwanted son to an unmarried couple in Milan, Italy. His mother, Chiara, already had several children and didn't want another one. When an herbal concoction failed to produce the desired abortion, she was gratified to give birth to a baby so ill that he was not expected to survive. Much to his mother's chagrin, the baby pulled through after taking a bath in red wine. So began the low-probability life of the first person to develop a theory of probability.

As our hero grew up, his lawyer father, Fazio, used him as a book carrier and mobile reading desk, weighing down the five-year-old with piles of heavy books and kicking him through the streets. Only when the boy became seriously ill at age eight did his father repent and have him baptized, giving him the name Gerolamo Cardano. Cardano was a curious Renaissance genius: physician, mathematician, gambler, mechanical engineer, and founder of probability theory. As Gerolamo grew up, he accompanied his father on visits to many of the lawyer's clients, whom Fazio consulted in geometry and law. When Gerolamo was thirteen, his father took him to meet the great Leonardo da Vinci. The boy had a voracious appetite for knowledge, learning Latin and geometry, and there was nothing that did not interest him, from witchcraft and horoscopes to the construction of spider webs and the circulation of the blood. He was well on his way to becoming a scholar and making a name for himself. Unfortunately, despite his great promise, his father refused to send him away for further education.

One day, when Fazio struck his wife in a fit of rage, she hit her head on a table. Fazio regretted his violent act immediately, but Chiara, having grown fond of her unwanted son, Gerolamo, milked the incident for all it was worth. Fainting repeatedly and crying out to her sister who had witnessed the event, she made Fazio promise to let her son attend college. Reluctantly agreeing to this blackmail, Fazio suggested law as it was a lucrative field of study. There was a stipend to be had, and Gerolamo's father, like many fathers, was eager to get his son an education without having to pay the tuition. Gerolamo, however, did not care about law. He wanted to be a medical doctor, but his father refused to pay for such an expensive course of study. Gerolamo had to find the money somewhere else.

He found it in gambling. Gerolamo preferred playing dice, because he had a natural sense for its probabilities. He did not cheat (a practice not recommended in a time when cheaters often found themselves hanging from the rafters), but he knew how to place bets. Before long, he had saved enough money to pursue medical studies at the renowned University of Padua. After some difficult years, including his annoying his fellow physicians by writing a dissertation about their poor practices, Cardano became a successful physician and chairman of the medical faculty in Padua. He wrote numerous books about medicine and mathematics, especially algebra. And he never forgot his gambling days. Wanting to share his experiences, he wrote the first theory of gambling, *Liber De Ludo Aleae* ("The book on games of chance").

Although Cardano's book was not published until a hundred years after it was written, it was a landmark work, introducing the fundamental idea of calculating probabilities: If you wanted to know the probability of a certain event out of all possible events, count the number of ways the event could occur, and divide it by the number of all possible events. This method of calculating assumed that all events were equally likely. Here is an example: What is the probability that you will roll a sum of 5 with two dice? There are 6 × 6, or 36, ways you can roll two dice: (1, 1), (1, 2), . . . (6, 5), (6, 6). How many ways are there to get a sum of 5? Count them up: (1, 4), (2, 3), (3, 2), (4, 1). That's four ways. Thus, the probability to get a sum of 5 with two rolls of a die is $4/36 = 1/9$. With some work, you can even work out the probability of a royal flush. Choosing five cards at random out of fifty-two cards provides 2,598,960 different possible hands. Only 4 of these will be royal flushes (one for each suit). Divide 4 by 2,598,960, and you get 1/649,740. You don't even have to play poker to figure this out.

Cardano makes an interesting hero for our story. As a physician, he understood the chance occurrences that play a role in people's lives. His life was a jumble of random events. He founded probability theory, invented a method to writing secret messages, and even had a connection to my current hometown of Detroit. He was the inventor of the Cardan shaft, or universal joint, still used in automobiles (and initially designed for a water-pumping system). Sadly, randomness finally got the better of him when a string of bad luck landed him in jail and finally in the poorhouse.

After Cardano, the mathematical treatment of games of chance became commonplace in the seventeenth and eighteenth centuries. Rich (and apparently quite bored) aristocrats sponsored mathematicians to figure the odds of various games. In one such case, the mathematicians Blaise Pascal (1623–1662) and Pierre de Fermat (1601–1665) were commissioned by the Chevalier de Méré to solve the *problem of points*: How should wagers be divided if a game is interrupted too early? Here is the scene: The Chevalier de Méré challenges the Comte de Dubois (a fictitious scenario) to a simple game of dice, and the winner is whoever throws the first 5 sixes. After ten minutes of play, the count is suddenly summoned to meet the king in Versailles. So far, he has thrown 3 sixes, and the knight 4. Neither of them trusts the other. But can the wager of sixteen pieces of gold be fairly divided? Clearly, the wager had to be distributed according to the

Row	
0	1
1	1 1
2	1 2 1
3	1 3 3 1
4	**1 4 6 4 1**
5	1 5 10 10 5 1

FIGURE 2.1. Pascal's triangle. Each number is the sum of the two numbers directly above it. For example, the number 6 in the fourth row is the sum of the two 3's directly above it (arrows). The numbers represent how many ways you can choose *k* (column) items out of a total of *n* (row) possibilities. This triangle was used by Pascal to solve the problem of points. The fourth row, which is discussed in the text, is highlighted.

probability that either player could still win the game. Pascal decided that this problem was too difficult to solve by himself. He contacted the renowned amateur mathematician Fermat to help him out. Fermat and Pascal corresponded about this problem for a while, until Fermat found a rather tedious way to solve it using Cardano's simple rule. This inspired Pascal to improve Fermat's result by devising a general formula, based on a triangle of numbers, now called Pascal's triangle.

The numbers in Pascal's triangle provide the number of ways you can choose a certain number of items (let's say *k* items) out of *n* available items. An example is the lottery. In a lottery, we pick, say, 5 numbers out of a possible 56, as in the MegaMillion game. In how many ways could we do that? A lot! Let's first try a simpler example: How many ways are there to select 2 items out of 4 available items A, B, C, and D? To find out, we go to the fourth row in Pascal's triangle (Figure 2.1). This row contains the numbers 1, 4, 6, 4, and 1. These numbers tell us how many ways there are to pick *k* items out of 4 available items (if we had 5 available items, we would need to look at row 5 of the triangle, and so on). The numbers 1, 4, 6, 4, and 1 correspond, from left to right, to picking 0, 1, 2, 3 or 4 items out of 4 available items.

Let's go step by step: As we move from left to right along the fourth row, the first number in the fourth row is a 1. This number tells us in how many ways we can select *zero* items out of 4. *Not* selecting an item can only be done in exactly one way (mathematics can be strange). The next number in the row is a 4; it tells in how many ways we can select *1* item out of the 4. Since there are 4 possible items, we have 4 ways to pick 1 of them (i.e., we can pick either A, B, C, or D). It gets interesting (and less obvious) when we pick more than 1 item out of 4. How many ways are there to pick *2* items out of 4? Looking at the next number in the fourth row, there should be 6 different ways to do this. And indeed there are: (A, B), (A, C), (A, D), (B, C), (B, D), and (C, D) (we are not allowing picking the same letter twice).

Now, to find out how many ways there are to pick 5 numbers out of 56 would require us to continue the triangle all the way down to the 56th row. You are invited to try, but you will quickly realize that it is a difficult task. The numbers grow quite large. Fortunately, there is a formula to calculate *any* entry in Pascal's triangle, without having to draw the triangle. It is called the *binomial coefficient*, given by $n!/((n-k)!\,k!)$, where n is the number of items, k is the number of possibilities, and $n! = 1 \cdot 2 \cdot \ldots \cdot n$. The expression $n!$ is called the factorial of n. For example, the factorial of 3 is $3! = 1 \cdot 2 \cdot 3 = 6$. For our lottery problem, we find that there are 3,819,816 ways to select 5 numbers out of 56. No wonder I haven't won the lottery yet.

How do the binomial coefficients help to solve the problem of points? Pascal and Fermat realized that you need to figure out in how many ways each person could still win the game. If the Chevalier de Méré needs only two games to win, and the Comte de Dubois three, what is the maximum number of rounds they need to play until one of the men is the winner? The answer is four rounds (or $2 + 3 - 1 = 4$). Why? If the knight wins none or only one game, then the count must have won at least three and is the winner. If the knight wins two or more rounds, the count must have won less than his needed three, and loses the game. Either way, one of them will be the winner.

Now that we know they need to play four more rounds to have a winner, we only have to calculate in how many ways the knight can pick his two wins out of the four rounds. And that is the same problem we just solved: He has six different ways to win (think of our items A, B, C, and D as labels for the four rounds they need to play). If he wins *more* than two rounds, he also wins the game, so we have to consider those possibilities as well. If he wins *three* rounds, he has four ways of doing so (see Pascal's triangle), and if he wins all four, there is only one way to achieve this feat. In total, he therefore has $6 + 4 + 1 = 11$ ways of winning the game. The count, by contrast, only has $4 + 1 = 5$ different ways of winning the game (picking three or four wins out of four rounds). Therefore, if they stop the game before the last four rounds, the ratio of the payout that each of them should receive is 11/5 in favor of the knight (who has more ways of winning the game than the count). If there are 16 pieces of gold left, the knight should get 11 and the count 5.

The Science of Ignorance

Around the time Pascal and Fermat solved the problem of points, a salesman of buttons and ribbons, named John Graunt (1620–1674), noticed an interesting pattern in the mortality rolls of London. It seemed that the number of deaths was always about the same every year, even though there were many causes of death and the exact circumstances of each death were unique. When he looked at a large-enough sample—provided by the city of London—Graunt found that individual differences became irrelevant and general patterns emerged. The science of statistics was born.

Statistics has been called the theory of ignorance. It's an apt description. Statistics is what we do when we face complex situations with too many influencing factors, when we are ignorant of the underlying causes of events, and when we cannot calculate a priori probabilities. In many situations, from the motion of atoms to the value of stocks, patterns emerge when we average over a large number of events—patterns not obvious from looking at individual events. Statistics provides the clues to understanding the underlying regularities or the emergence of new phenomena arising from the interaction of many parts.

The work of Graunt led to the first *life tables*, which gave the probability that a newborn would end up living to a certain age. This was the kind of information life insurance companies needed to make money: If you insured enough people and knew your life tables, you could charge people enough money to make sure you ended up in the black, even if occasionally someone died before his or her time. Life insurance became well-informed gambling, with probabilities taken from real life. In his book *The Drunkard's Walk*, Leonard Mlodinow reproduces Graunt's life table for London in 1662. In the late 1600s, 60 percent of all Londoners died before their sixteenth birthday. Such an awful statistic makes modern-day Afghanistan look like paradise. There, the death rate of 60 percent is close to age sixty. By comparison, the 60 percent death rate in Japan is around ninety years old.

Although statistics emerged from the need to quantify economic and sociological data, it was soon recognized that this new science could benefit the hard sciences as well. Repeated measurements of the same phenomenon, especially in astronomy, were observed in the eighteenth century to follow a law of errors: Errors seemed to obey a universal distribution. However, it was difficult to find the correct mathematical function that would fit the error distributions. After all, every set of measurements only fit the distribution approximately, and the approximation only became good enough to allow guessing the right function after a huge number of measurements. After several false guesses by various eminent mathematicians, the German mathematician Carl Friedrich Gauss (1777–1855), using some of his astronomical data, recognized that the so-called normal distribution seemed to fit the bill.

The normal distribution had been under mathematician's noses all along. The French mathematician and gambling theorist Abraham de Moivre (1667–1754), in his 1733 book, *The Doctrine of Chances*, had published a formula that extended Pascal's triangle to very large numbers of trials, much larger than could be practically obtained by Pascal's method. In the limit of large numbers, Pascal's triangle could be approximated by a formula describing a curve that looked like a bell. This bell curve, or normal distribution, is what Gauss found in errors of astronomical data. The French mathematician Laplace picked up where Gauss left off and proved that any measurement that depends on a number of random influences tends to have errors that follow the normal distribution. Today, Laplace's central limit theorem is a key part of statistics, which can predict distributions as varied as people's heights or masses of stars.

The use of the normal distribution in statistics was perfected by the Belgian scientist Adolphe Quetelet (1796–1874), who subjected everything he could get his hands on to statistical analysis: chest sizes of sailors, heights of men and women, murders committed with various weapons, drunkenness, and marriages. Wherever he looked, he found the normal distribution. And when he did not find it, he knew that something had gone awry. When, for example, he found the height of French military conscripts strangely deficient at the low end of the scale, he realized that many short men of military age had lied about their height to get out of serving. The minimum size for military service was 157 centimeters (5 feet 2 inches). If you were 157.5 centimeters, why not buckle your knees a little bit during measurement and escape the dreaded military service?

Quetelet's work illustrated how the error law, the normal distribution, and the central limit theorem governed almost everything. His books were eagerly read not only by future sociologists, but also by future physicists and biologists. If statistics was useful in economics, medicine, and astronomy, why not in other areas as well? As we will see in Chapter 3, nineteenth-century physicists like James Clerk Maxwell and Ludwig Boltzmann used Quetelet's work to develop the statistics of atoms and molecules.

Mathematician Francis Galton, Charles Darwin's cousin, was one of the first to apply Quetelet's ideas to a wide range of biological phenomena. He found that the normal distribution governed almost every measurement of an organism: heights, masses of organs, circumferences of limbs. One of his most important discoveries was *regression toward the mean*: Galton found that any offspring of a parent who was at the outer ranges of a distribution, for example, a very short man or a very tall woman, generally tended to "regress" toward the mean of the distribution. In other words, the son of an exceptionally short man tended to be taller than his father, and the daughter of an extremely tall mother tended to be shorter than her mother. Mozart's children were not geniuses like their father; neither were Einstein's. And parents of below-average intelligence often have smarter children. In the long run, we all tend toward the average. In some sense, this is a good thing. Genius is unpredictable—which makes it all the more puzzling that Galton became one of the founders of eugenics, the idea that selective breeding of humans could improve humanity. Beyond the obvious human rights issues with this awful idea, Galton's own "regression to the mean" suggested that the prospect of success would have been highly questionable. "Breeding" two highly intelligent humans would never guarantee that their offspring would be more intelligent than, or even as intelligent as, their parents. According to Galton's own "regression towards the mean," the best bet may be on "less intelligent."

Despite his tragically misguided ideas, Galton made other important contributions to statistics, such as the *coefficient of correlation*. This statistic measured how two different variables were statistically linked, or correlated. One of the main statistical tools of modern biology and medicine, the coefficient of correlation is mathematically sound but has to be used with care. Just because stork numbers and baby births may be correlated over a few years does not mean that storks bring babies. A correlation is a hint of a possible connection, not a proof.

With Galton's and Quetelet's ideas, statistics flourished in the nineteenth century. It became an integral part of biology, and through his use of statistics, Mendel discovered the laws of heredity. The scene was now set to recognize randomness as an important player in the story of life.

Randomness and Life: Three Views

The question of chance versus necessity occupied the human mind for thousands of years. For much of this time, philosophers, theologians, and scientists denied randomness any meaningful role in nature, with atomism being the most obvious victim of randomness phobia. However, with scientific advances in the late nineteenth and early twentieth centuries, this stance became more and more difficult to maintain. Randomness reared its ugly head first in biology, through Quetelet and Galton's work, and finally in physics. The final battles over randomness played out in debates over the existence of atoms, the emergence of statistical and quantum mechanics, and new insights into molecular evolution and the role of mutations.

The randomness debate involved several famous protagonists. Nineteenth-century Austrian physicist Ludwig Boltzmann, cofounder of statistical physics, fought for the existence of atoms, which were still disbelieved more than two thousand years after Democritus and Epicurus. Twenty years later, Albert Einstein proved the existence of atoms in his famous papers on Brownian motion, but was unhappy with the implication that the new science of quantum mechanics was at heart built on randomness. These debates also continued in biology. Ever since Quetelet and Galton had shown aspects of physiology to be governed by statistics, and ever since the theory of evolution had introduced randomness as a driver of novelty in the development of life, pitched battles were fought over the role of chance in the history of life. Three of the most prominent combatants included D'Arcy Wentworth Thompson (1860–1948), Pierre Teilhard de Chardin (1881–1955), and Jacques Monod (1910–1976).

An English mathematician and biologist, Thompson was best known for his 1917 masterpiece *On Growth and Form*, a book filled with astonishing insight (and many fascinating diagrams) of how mathematical principles guide the shapes and forces of living organisms. Thompson did not believe in teleological life forces. It seemed frivolous to him to invoke nebulous reasons, when a mathematical or physical explanation would do. He saw a continuity of complexity acting throughout all of nature and therefore no unbridgeable chasm between the living and the dead. "The search for differences . . . between the phenomena of organic and inorganic, of animate and inanimate things, has occupied many men's minds, while the search for community of principles or essential similitudes has been pursued by few. . . . Cell and tissue, shell and bone, leaf and flower, are so many portions of matter, and it is in obedience to the laws of physics that their particles have been moved, molded and conformed."

In all of this, Thompson was humble—he allowed the possibility that not all phenomena of life could be explained through physical laws alone. But he felt that too often, biologists of his time surrendered too early, and that many phenomena of life *could* be explained, if scientists were given sufficient time to gain insight into the involved complexities. In the battle between purpose and mechanism, he understood the usefulness of invoking final causes, but realized that one cannot stop there, but must find the physical reasons for how structures arise. "In Aristotle's parable, the house is there that men can live in it; but it is also there because the builders have laid one stone upon another." For Thompson, a full explanation needed to contain an explanation of both why a structure was there and how it was constructed.

On the other hand, Thompson, while fond of mechanistic explanations, did not support the theory of evolution. For him, explanations should be explanations of necessity—chance was to play no role. He also did not favor theological hand-waving: "How easy it is, and how vain, to survey the operations of Nature and idly refer her wondrous works to chance or accident, or to the immediate interposition of God." Invoking chance, God, or any extraneous life principle when met with ignorance was a cheap trick, according to Thompson, designed to keep us from doing the hard work of finding the true causes.

Where Thompson had polite disdain for final causes, Chardin, a French Jesuit priest, paleontologist, and anthropologist, celebrated them—envisioning even atoms and molecules as bound to a higher purpose. Chardin made the reconciliation of science and religion his life's work. As a scientist, he knew that the theory of evolution was the best explanation for the development of the living world, and he became an enthusiastic champion of evolution, although with a twist: In his masterpiece *The Phenomenon of Man* (written in the 1930s, but published in 1955), Chardin envisioned evolution as an upward motion toward more complex and sophisticated forms of life, which would ultimately culminate in a single, universal mind, which he equated with God. According to Chardin, evolution was guided by a mysterious psychic energy, an energy not yet measured or discovered, but nevertheless evident from the progress seen in the evolution of our universe. Mind was primary, pulling matter along in its wake. Chardin's philosophy was to give "primacy to the psychic and to thought in the stuff of the universe." Voltaire had made fun of such ideas two hundred years earlier, but now a better understanding of the awe-inspiring history of evolution, and the need to define humanity's place, made such animistic philosophies fashionable again.

Chardin's philosophy seemed to offer a way to have your cake and eat it too: You could embrace evolution and all the findings of science, and at the same time, believe in a higher guiding force, although somewhat invisibly. Yet, despite his compromising stance on the big questions of our existence, few were happy with his ideas. The Church felt they were too far removed from traditional theology, and scientists didn't appreciate the addition of unknown forces just to satisfy a need to place humanity in the center of all being.

One of the scientists who regarded Chardin's ideas with disdain was the French biochemist and Nobel Prize–winner Jacques Monod. In "Vitalisms and Animisms," a chapter in his book *Chance and Necessity*, Monod critiqued Chardin's logic as "hazy" and his style as "laborious." But most of all, he was "struck by the intellectual spinelessness of [his] philosophy." There seemed to be a "willingness to conciliate at any price, to come to any compromise."

Yet, even for Monod, the problem of how to reconcile the blind motions of atoms and their lifeless laws with the complexity of life was a deep, central mystery. Allergic to any vitalistic or animistic explanations and unable to see how physical laws by themselves could lead to the complexity of life (which was Thompson's idea), he resolved the dilemma by placing randomness front and center. The origin of life, according to Monod, was an incredibly improbable event. Once it happened, evolution took over, infused with a healthy dose of randomness. While both Thompson and Monod rejected the introduction of guiding principles, such as Chardin's spiritual energy, they greatly differed on the roles of chance and necessity.

The problems of how life came to be, how life operates (metabolism, growth), and the history of evolution can be answered in many ways. However, we could organize most answers to these questions along two dimensions. There is the dichotomy of "mere physics" versus higher forces (life forces, the soul) along one dimension, and then the question of chance versus necessity along the other. Figure 2.2 shows schematically how we could visualize these various views. Why are there so many views on the origin and nature of life? Undeniably, it is difficult to see how simple physical laws can lead to life's complexity, but at the same time, scientists have repeatedly found that yesterday's ignorance about certain biological phenomena have turned into today's knowledge—knowledge based on those same simple physical laws. Thus we should heed both Monod's and Thompson's warning: When we feel the need to invoke extraneous principles to assuage our ignorance, it is wiser to hold off. Rather, we should continue our search for explanations within known science. So far, this has served us astonishingly well. Chardin knew that the "connection of physics and biology" had to lie in the cell. However, he believed that the cell was "still a closed book" and "an impregnable fortress." Since Chardin's time, much of this fortress has been penetrated, the depth of our explanations has greatly increased, and the cell is not quite as enigmatic as it used to be. And in all of this, physics and chemistry have been our guides.

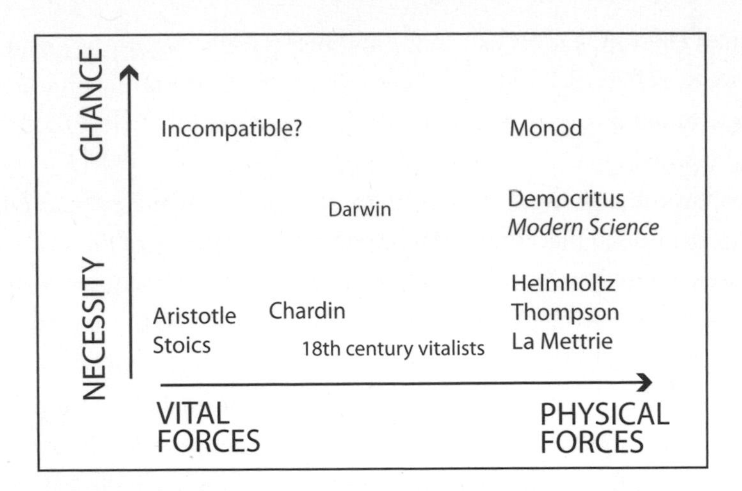

FIGURE 2.2. A simplified view of how various scientists and philosophers have explained life with respect to chance versus necessity and vital forces versus physical forces.

Variation and Atomic Physics

Vitalism was discredited by physics by the end of the nineteenth century, and the *Origin of Species* had discredited purpose—but many scientists could not yet accept the idea that randomness might play an important role. Darwin's theory provided a plausible explanation for the variety of life forms and the biological history of our planet, an explanation that did not involve purpose or teleology. But physics and evolution made strange bedfellows. Physics of the nineteenth century was based on natural laws. It was based on necessity. By contrast, evolution needed variation and novelty—chance—to function. How could the two be reconciled?

Reconciliation came slowly and happened through the express help of physicists. By the end of the 1800s, the holistic laws of thermodynamics—the laws that describe the behavior of matter using properties such as pressure, volume, or temperature—were reduced to a more fundamental theory. This theory explained these holistic properties in terms of the motions of atoms. The new theory, which we will explore in more detail in Chapter 3, was at first called kinetic theory, but as it grew and encompassed more and more phenomena, it became known as statistical mechanics. Randomness became an accepted part of physics, and tamed by statistical averaging over large numbers, the random motions of atoms could now be described by well-defined probability distributions.

Early in the twentieth century, the study of atoms and light led to another theory that explicitly included the concepts of chance and probability: quantum mechanics. The iron-clad model of necessity, classical physics, was now replaced by a fundamentally statistical picture of nature—a picture in which we could never state with certainty where a particle would go or how much energy it had. All we could calculate were probabilities. Quantum mechanics arose from a need to explain startling new experimental results. For example, throughout the late 1800s into the early 1900s, experiments revealed a plethora of new and mysterious radiations: X-rays; cathode rays; and alpha, beta, and gamma radiation. The study of these new types of radiation provided impetus for the new science of the quantum. And by the 1920s, these mysterious rays would also prompt a sea change in biology: Physicists were starting to study the effect of radiation on biological matter.

Chromosomes, bundles of DNA, were discovered in 1882 by German biologist Walther Flemming (1843–1905) and others, but their significance was not immediately clear. Although they were duplicated during cell division, the part they played in heredity was not recognized, because Mendel's work had fallen into obscurity. By 1900, however, his work had been rediscovered and the German biologist Theodor Boveri (1862–1915) and his American counterpart Walther Sutton (1877–1916) made the connection between chromosomes and Mendel's hereditary traits. By 1909, the American embryologist Thomas Hunt Morgan (1866–1945) had begun his famous genetic experiments on the fruit fly *Drosophila*, a fast-reproducing animal. The momentum of biological research now shifted across the Atlantic. Morgan discovered that not all traits were independent, as Mendel had thought, but that there were various degrees of linkages. This suggested that traits were contained in some kind of linear arrangements on chromosomes, with nearby traits more likely to be inherited together. The mixing of traits was assigned to a crossing-over of linear molecules. During the crossing-over process, the progeny received a mixture of the genes from both parents. If two traits were located close to each other on the hereditary molecule, it was less likely that they would be separately inherited during the reshuffling. However, crossing-over only explained some aspects of variations in populations. It could not explain how brand new traits could arise, but these new traits were needed for evolution.

By 1920, through the tireless work of several pioneers in genetics, it became clear that hereditary information was lined up along linear molecules, wound into chromosomes. The proof came from the new science of radioactivity, and with the understanding of radiation came a new idea on how novelty and variation were introduced into a species beyond a mere reshuffling of existing traits. One of these pioneers was the American geneticist Hermann Joseph Muller (1890–1967), one of Morgan's Ph.D. students. Muller became interested in the effects of X-rays and radioactivity on the mutation rates of fruit flies. He had been studying mutations; these rare and significant changes in the hereditary material are generally harmful. He was hoping that X-rays could induce mutations in a controlled manner. After three years of false starts (the X-rays sterilized the fruit flies, and they produced no offspring he could study), a breakthrough came in 1926. By controlling the dose of X-rays, Muller was able to find a direct relationship between X-ray dose and the probability of mutation. This work established that radiation increased the probability that new genetic traits would be created in a species—chance was finally coming into its own.

Muller obtained definitive results in 1932, working in Berlin with Russian geneticist Nikolai Timoféeff (1900–1981), but the molecular nature of the hereditary substance remained a mystery. At this point, a young atomic physicist joined Timoféeff's lab. Max Delbrück (1906–1981) was able to explain Muller and Timoféeff's data theoretically using his knowledge of atomic physics. Although the nature of the genetic substance was unknown, Delbrück argued that if we assumed the genetic substance to be a molecule, it should be subject to the laws of atomic physics and of thermodynamics. In the now famous green pamphlet of 1935, published in the obscure *Transactions of the Scientific Society of Göttingen*, Delbrück, Timoféeff, and Karl Zimmer (1911–1988) presented data on the dependence of mutation rates on temperature or X-ray dose. The results were clearly compatible with current knowledge of atomic and thermal physics.

Thus, a remarkable story unfolded throughout the first half of the twentieth century: Previously mysterious *biological* processes, such as heredity and variation, became connected to measurable *physical* entities.

By contrast, Helmholtz's achievement had been essentially restrictive—it subtracted vital forces from the list of possible explanations. However, Helmholtz and his fellow nineteenth-century scientists could not explain how the business of life was conducted. This business was conducted on the molecular scale, which had been inaccessible to nineteenth-century science. For the first time, through the work of Muller, Timoféeff, Delbrück, and others, some of the deepest mysteries of life were connected to physical, *molecular* entities. Molecular biology was born.

What Is Life?

As an alumnus of Johns Hopkins University, I can appreciate an anecdote I found in Walther J. Moore's biography of Erwin Schrödinger (1887–1961), 1933 Nobel laureate and founder of wave mechanics—the most widely used formalism for quantum mechanical calculations. Schrödinger, a scientific refugee from Nazi Austria, was offered a position at Johns Hopkins. In accordance with time-honored tradition, faculty members of the host university wanted their distinguished guest to have a good time. In true Baltimore fashion, they gathered at a seafood restaurant to sample the famous Maryland crabs. I am not sure if they already had Old Bay seasoning at the time, but Schrödinger enjoyed his seafood very much and felt a nice glass of beer would go very well with it. Unfortunately, it was the height of prohibition. The hosts apologized, but no beer was to be had. Schrödinger decided to take a position in Dublin, Ireland, instead.

Delbrück's work would have lingered in the obscure journal in which it was published had it not been brought to Schrödinger's attention in the early 1940s by another émigré from Nazi rule, Paul Peter Ewald (1888–1985), a pioneer in X-ray physics. Schrödinger was fascinated when he first read the green pamphlet, which was titled "On the Nature of the Gene Mutation and the Gene Structure." For Schrödinger, the green pamphlet by Delbrück, Timoféeff, and Zimmer was a revelation.

Schrödinger worked the technical paper into a series of inspiring lectures at Trinity College, Dublin, attended by over four hundred listeners and, later, into his famous book *What Is Life?* In the book, he placed Delbrück and his colleagues' findings into the context of contemporary science and made daring speculations based on the rather more careful conclusions of these scientists. Although many of Schrödinger's speculations were wrong, the book provided an inspiration to many physicists entering biology.

The book was quite short (only ninety pages in the edition I own), but Schrödinger touched on many of the puzzling aspects of life, especially the nature, size, and surprising stability of the hereditary substance. Schrödinger was familiar with rough estimates of the size of the hereditary substance from microscopic observations of chromosomes, which had placed the size of a gene at about 30 nanometers cubed—still large for a molecule. Schrödinger wanted a closer estimate. Delbrück, Timoféeff, and Zimmer had estimated that if X-rays were strong enough to ionize about one in every thousand atoms, then mutations would occur with near certainty. Assuming that gene mutations were due to atomic changes in a gene-carrying molecule, Schrödinger took Delbrück's work one step further: If ionizing one in a thousand atoms caused a mutation with (almost) certainty, then the size of a gene had to be about one thousand atoms, which was about 3 nanometers cubed.

This conclusion, which Delbrück and his collaborators had wisely avoided, did not make much sense, even to contemporary molecular biologists. Mutation happens due to the creation of so-called radicals (molecules with a missing electron), which can diffuse over much larger distances than 3 nanometers. Moreover, today we know that there are molecular machines, so-called repair enzymes, that can repair the genetic material (which we now know to be DNA). Thus, mutations are really the result of chemical damage, which is subsequently not repaired correctly by the cellular machinery. Nevertheless, Schrödinger's estimate, although based on false premises, spurred the imagination. What could fit into a 3-nanometer cube? What kind of molecular units would consist of one thousand atoms?

Using his estimate of the size of a gene, Schrödinger wondered how such an assembly of atoms could be stable. Molecules in a living body are subject to violent thermal motion—at the elevated temperatures of a living body, atoms rattle, shake, and bump into each other at high speeds. Only a very stable chemical bond could survive such abuse. Schrödinger became convinced that genes must be molecules. He envisioned the genetic material to be like a crystal, but with one unexpected condition. To hold the complex information needed to operate a cell, the crystal had to be *aperiodic*, that is, nonrepetitive. Real crystals are quite boring on an atomic scale—they are repetitions of the same atomic arrangement over and over. Such a repetitive arrangement cannot contain much information. It is like writing a book, but you are only allowed to use one letter: *eeeeeeeeeeeee*. To convey information, you need different letters, which can be arranged into sentences, such as "The cow jumped over the moon." You need an aperiodic sequence of letters. Schrödinger believed that the letters of the genes were written in the language of atoms and molecules. Here, Schrödinger was closer to the mark than with his estimate of the size of the letters, although this idea was not original with him. We now know that the genetic material is not an aperiodic crystal, but an aperiodic polymer: a floppy, long, linear molecule, called DNA.

Schrödinger's puzzlement over how the molecules in our cells escape thermal motion led him to conclude that everything in our cells was made stable by strong chemical bonds. For him, thermal motion was the enemy, to be overcome by fortifying the bonds of our microscopic nature. As we will see throughout this book, Schrödinger was fundamentally wrong on this point. There are no solids in our cells. Everything is squishy and moving. Far from being the enemy, thermal motion is the key to the activity in our cells.

Schrödinger also commented on the value of statistical mechanics, the science of averaging large numbers of randomly moving molecules to arrive at precise macroscopic laws. An example is the ideal gas law, a law that relates the density, pressure, and temperature of a gas. This law emerges from averaging vast numbers of gas molecules. Schrödinger called this process "order from disorder." In biology, by contrast, Schrödinger saw a different class of laws at work, laws that made "order from order." Undoubtedly, that is what living organisms do, but deep down, they still have to contend with disorder and must first make order from this underlying chaos. Schrödinger could not see how this was possible. The numbers of atoms in life's molecules seemed to be much too small, and expected random changes (or "fluctuations") much too large. In a stunning reversal of Helmholtz's insights, Schrödinger claimed that biology had to encompass new laws of physics not previously seen in inanimate matter. According to him, statistical mechanics could not, by itself, explain living matter. Instead, the "most striking feature" of life was that it seemed to be based on an "order-from-order principle" rather than an order-from-disorder principle as in statistical physics. Schrödinger's solution was to imagine life as clockwork, in a throwback to La Mettrie two hundred years previously. However, he admitted that the idea of life as clockwork had to be taken "with a very big grain of salt." A big grain, indeed.

We are now close to a solution of Schrödinger's conundrum. A living organism is not based on a solid. It is not clockwork. And statistical mechanics can teach us a lot about how it works.

The Rules of the Game

As we've seen in this chapter, randomness is here to stay. Far from the destructive force it has been made out to be through the millennia, it is good for us—or at least good for life as a whole. In the chapters that follow, we will learn how randomness is part of every aspect of life—even in the simple act of lifting an arm or converting food into motion.

How can we visualize the relationship between necessity, the laws of nature, and randomness? One way was suggested by biochemist Manfred Eigen in his 1975 book *The Laws of the Game: How the Principles of Nature Govern Chance*. Eigen won a Nobel Prize for the study of ultrafast chemical reactions and realized that the interplay of necessity and chance resembles games. Different games can represent different phenomena we may encounter in the principles that govern life: chemical reactions, population growth, the regulation of enzymes in cells, or evolution. A good game combines elements of necessity (it must have rules), chance (there must be surprise), and sufficient complexity. Chess has simple rules, but the totality of all chess games ever played and the variety of chess strategies used show that chess is a game full of subtle complexity. Games are models of emergence—the appearance of unexpected features arising from the interactions of many different parts, rules of the game, chance, and space and time. Life can best be understood as a game of chance—played on the chessboard of space and time with the rules supplied by physics and mathematics. To gain a physicist's understanding of life, we need to begin with the rules the game of life obeys. To start, let us learn about the kind of games atoms play.

3
The Entropy of a Late-Night Robber

> It has been [our] principal indeavour to enlarge and strengthen the Senses . . . by . . . outward Instruments. . . . By this means [we] find . . . that those effects of Bodies, which have been commonly attributed to Qualities, and those confess'd to be occult, are perform'd by the small Machines of Nature, which are not to be discern'd without these helps.
>
> —ROBERT HOOKE, *MICROGRAPHIA*

> So nat'ralists observe, a flea
> Hath smaller fleas that on him prey,
> And these have smaller fleas that bite 'em,
> And so proceed ad infinitum.
>
> —JONATHAN SWIFT

WHEN ROBERT HOOKE PEERED THROUGH HIS PRIMITIVE microscope, he found a new world of tiny "Machines of Nature," from dimples on poppy seeds to the sting of a bee. The "machines" he saw through his microscope in the late 1600s were just the beginning: As microscopes improved, all of Hooke's machines were found to be made of cells, which themselves were entire factories of even smaller machines, each made of smaller parts yet—all the way down to atoms and molecules.

It became clear that living things, while immensely complex, were made of the same stuff as the rest of nature.

How do atoms and molecules assemble into a flower or a human? Where do we cross the threshold from lifeless atoms and molecules to living organisms? What makes an object alive? These hard questions puzzled scientists and philosophers for millennia. Yet, we may be the first generation to glimpse answers to these questions. To understand these answers, we must begin with the basic building blocks of nature: atoms and molecules. Atoms are tiny clumps of matter, so tiny, that it takes 300,000 carbon atoms to span the width of a single human hair. A humble *E. coli* bacterium is only one-quadrillionth (10^{-15}) the mass of a human, and yet it contains 100,000 billion atoms. Molecules are assemblies of atoms bound together by strong electrical bonds. Molecules can contain as few as two atoms and as many as tens of thousands.

Atoms and molecules are restless. Democritus, Epicurus, and their fellow atomists already understood this important point. In air, molecules of nitrogen, oxygen, carbon dioxide, and water vapor randomly swirl around, colliding at high speeds. Without noticing, we are continuously bombarded by supersonic gas molecules from the surrounding air. The calmness we see around us is an illusion. We are surrounded—no, immersed—in chaos. Yet from such chaos order can arise: On cold winter days, randomly swirling water molecules, high in the clouds, find each other and create beautiful, symmetric snowflakes. The world we see around us—the *macroscopic* world—is one of order and regularity. A book on a table does not jump suddenly; nor does it spontaneously burst into flames. Yet, seen at a very small scale, a book is a mass of atoms that rattle and shake, collide, and send each other hurling off into space. How can visible order and life's complexity arise from such chaos?

In the late 1800s, this question occupied physicists such as Ludwig Boltzmann in Austria, James Clerk Maxwell in Scotland, and Josiah Willard Gibbs in the United States. For them, the relatively simple example of a gas provided the perfect starting point. A macroscopic volume of gas follows simple laws that relate pressure, volume, and temperature, but how did these laws arise? To find the answers, these scientists turned to the new science of statistics, and invented *statistical mechanics*. This discipline applies statistics to the mechanics of atoms and molecules. In their thinking, if statistics can describe the height of a thousand men or the marriage age of a thousand women, it sure should be able to describe the behavior of a billion billion atoms. In everyday life, we use statistics to calculate such figures as average income, IQ (which is defined as a standard deviation from average intelligence), and income distributions. Similarly, physicists discovered how to calculate averages, deviations, and distributions of the speeds and energies of atoms. Although atoms move randomly, their collisions conform to physical rules. Maxwell and Boltzmann showed that the distribution of speeds in a gas was just a normal distribution—originally derived to generalize Pascal's gambling formula. Applying statistics to the chaos of atoms and molecules, they found that averaged over time and space, the randomness of atomic motion gives way to order and regularity.

Life is based on molecules. These molecules are subject to the underlying chaos of the molecular storm—which at first glance seems to be a destructive force. How can life survive and possibly benefit from this chaos? This was Schrödinger's famous question. Schrödinger saw a contradiction between the chaos of atoms and the structure of life. But today we know that the chaotic motions of atoms and molecules—controlled by life's intricate structure—give rise to life's activity. There is no contradiction. Life emerges from the random motions of atoms, and statistical mechanics can capture the essence of this emergence.

Money on the Rocks

When I was a graduate student in Baltimore in the early 1990s, I had the unfortunate experience of being robbed at gunpoint. All I had on me was ten dollars, so it was not a big loss, but it was an upsetting experience nevertheless. Money is what makes the world go round, but it also makes people do unpleasant things. In physics and biology, we have a different currency that makes things happen—we call it energy. Money and energy have a lot in common. In a transaction, where one party gives money to another party, the total amount of money is conserved: The robber gained ten dollars, and I lost ten dollars. Is it possible, in the same transaction, for me to lose eight dollars and the robber to gain ten dollars? No. Money does not appear out of nowhere.

The same is true of energy. As Helmholtz had shown, energy conservation is the strictest law of nature. In "energy transactions," the energy *before* equals the energy *after* the transaction. Energy can be transferred from one object to another or converted to a different form, but energy is never gained or lost.

Imagine you are standing on the moon, in the absence of air friction, and you pick up a rock. As the rock rests in your outstretched hand, it has *gravitational energy* (stored in the attraction between the rock and the moon). When you drop the rock, it accelerates as it falls. Motion is associated with a form of energy called *kinetic energy*. Where does the energy for the rock's motion originate? It comes from the gravitational potential energy stored in the attraction between the rock and moon when you lift the rock off the ground. The falling rock "pays" for the kinetic energy (motion) by using up gravitational energy. Throughout the rock's fall, the *total* energy of the rock (gravitational *plus* kinetic) is always constant. Energy is conserved.

Suddenly, your rock hits the ground in a cloud of moon dust and stops moving. Having lost gravitational energy through the fall, it loses its kinetic energy as well. If energy conservation is true, where does the energy go on impact?

Throughout the eighteenth and nineteenth centuries, physicists studied what happened when moving objects were slowed down by impact or friction. The law of energy conservation was new, and situations like our falling rock presented a conundrum. This changed when Sir Benjamin Thompson, Count Rumford, studied the heat generated while boring a cannon from a cylinder of metal. Finding that motion (kinetic energy) could be continuously turned into heat, he therefore concluded that heat had to be a form of energy. Before this realization, heat was thought to be a fluid (called *caloric*) that flowed from hot to cold objects, a fluid that eventually ran out. But Rumford refuted this idea: "It is hardly necessary to add that anything which any insulated body can continue to furnish without limitation cannot possibly be a material substance; and it appears to me to be extremely difficult . . . to form any distinct idea of anything capable of being excited and communicated, in the manner the heat was being excited and communicated in these experiments, except it be motion." With the knowledge that heat could be created from motion, or kinetic energy, scientists wondered, "What kind of energy *is* heat?"

Matter is made of atoms, which are in perpetual motion. How do we know this? For a gas sealed in a container, an increase of temperature is always associated with an increase in pressure. Early work by Maxwell, Boltzmann, and others in the kinetic theory of gases explained this pressure increase by relating both temperature and pressure to the motion of atoms. In this view, pressure was the result of innumerable impacts by atoms with the walls of the container. The faster the atoms moved, the harder they hit the walls, and therefore, the greater the pressure. The atoms could be made to move faster if heat was added and temperature was increased. Temperature seemed to be related to the kinetic energy of the atoms in the gas. Kinetic theory neatly explained the macroscopic laws governing gases, but it presumed the existence of hypothetical, continually moving atoms—tiny objects no one had ever observed. Boltzmann, the Austrian father of statistical mechanics, suffered extreme ridicule for suggesting the existence of atoms. This added to his deep depression, which ended in his suicide in 1906.

Yet the idea that gases such as air were made of restless atoms was not a new idea. Atomic motion was indirectly discovered by the botanist Robert Brown in 1827. Like his successor, Charles Darwin, Brown made his mark as a naturalist serving on a British surveying expedition. During his voyage, Brown collected thousands of Australian plant specimens, many of them previously unknown species—only to lose most of them in a shipwreck. Nevertheless, he became a well-respected naturalist, who is credited with naming the cell nucleus. Despite his daring travels, physicists best remember Brown for a discovery he made in the safety of his own home. Brown observed that pollen grains suspended in air or liquid perform a jittery dance, as if pushed by an invisible, random force. Today we call this dance *Brownian motion*. In some sense, Brown really only rediscovered what Democritus had observed two thousand years earlier—the "motes in the air" that were "always in movement, even in complete calm."

When Brown discovered the random motion of pollen grains, he wondered if their motion had something to do with the fact that pollen grains were alive. He decided to study suspended dust particles of similar size. They, too, performed the same strange dance. Brown concluded that the random motion was not due to the pollen's being alive, but was due to some inherent motion in matter. The solution to the mystery came almost a hundred years later, when Albert Einstein proved that the random movement of much smaller particles caused the jittery dance of the pollen or dust grains. Dust and pollen grains are jostled around by random collisions with countless atoms. A pollen grain moves because of small temporary imbalances between the number of atoms hitting from one direction and the number hitting from the other. Einstein suggested experiments to prove his theory of Brownian motion. Using the most sophisticated microscopes of the early 1900s, the French physicist and Nobel laureate Jean Perrin (1870–1942) used Einstein's theory to prove once and for all that atoms exist and are always in motion. Boltzmann, just two years after his tragic death, was vindicated.

The tiny scale of atoms and molecules is dominated by continuous motion. Scientists call this continuous motion of atoms and molecules *thermal motion*. Thermal motion does not mean gently floating atoms: At room temperature, air molecules reach speeds in excess of the fastest jet airplane! If we were reduced to the size of a molecule, we would be bombarded by a *molecular storm*—a storm so fierce, it would make a hurricane look like a breeze. Yet, despite their stupendous speeds, molecules in the air do not get very far, because they frequently collide with each other. When this happens, the colliding molecules bounce like tiny billiard balls. The jittery dance that Brown observed and that Einstein explained is the result of this underlying tempest of colliding atoms.

The Mystery of the Missing Energy

Now we can return to the falling moon rock. If energy is supposed to be conserved, what happened to the rock's kinetic energy at impact? Where did the kinetic energy of the rock disappear to?

Atoms cannot move as freely in a solid as they do in a gas or liquid. But the atoms in a solid are still moving; they oscillate at high frequencies about a central position, like the head on a bobblehead doll. Physicists often think of solids as collections of atoms connected by springs, all wobbling about, while on average staying at particular positions. When the rock hit the ground, the kinetic energy of the rock did not disappear. Instead, the kinetic energy of the rock transferred to atoms in the rock and in the ground, making the atoms wobble more vigorously. As a result, the rock and the ground got warmer. The kinetic energy of the rock was converted into thermal energy or heat. Energy remained conserved.

The conservation of energy, coupled with the fact that heat is a type of energy, is called the *first law of thermodynamics* (we will meet the second law shortly). Thermodynamics is the science that deals with thermal energy (the word comes from the Greek *therme*, meaning "heat," and *dynamis*, meaning "power") and is the macroscopic "sister science" of statistical mechanics. Thermodynamics is what emerges when we average the random motions of atoms using the tools of statistical mechanics.

Now that we have solved the mystery of where the energy of the falling rock went on impact, let me ask a dumb question (science has advanced by asking a lot of these): Why don't rocks extract heat from the ground and jump up spontaneously? This would not violate energy conservation. The rock could take heat from the ground, making the ground cooler, and turn the extracted heat into kinetic energy. Yet, we never see this happen. Rocks don't spontaneously jump off the ground. Why not?

After our rock hit the ground, the atoms in the rock and the ground started to shake more violently. Both the rock and the ground became warmer. Atoms in solids are attached to other atoms, so if one atom shakes, neighboring atoms will soon shake as well. When an atom excites its neighbors, the atom loses some energy, which its neighbors gain. In turn, the atom's neighbors excite *their* neighbors, and the extra energy provided by the rock's impact is soon randomly distributed among an astronomical number of atoms in the rock and in the ground.

In the story of the late-night robber, the robber stole my money, he spent some of it, and the people who received money from him spent their money, too. Imagine that the robber stole a thousand pennies instead of a ten-dollar bill. After a while, one thousand people could potentially each have a penny of my money. It would be highly unlikely that my pennies would be spontaneously reunited, as this would require one thousand people (probably unacquainted) to go to the same merchant at the same time to spend their pennies. Similarly, it would be impossible for all the atoms in our rock to spontaneously concentrate their energy to make the rock jump. If we can believe that it is close to impossible for one thousand pennies to be spontaneously reunited, consider the immense number of atoms in the rock (something like a trillion trillion atoms!). This giant number of atoms would have to simultaneously push in the same direction for the rock to jump up from the ground. Yet, there is no master choreographer that tells the atoms in which direction to shake. They all shake randomly.

Consider the argument I have just made. I did not say that it is *impossible* for a rock to extract heat from the ground and spontaneously jump up from the ground. The word *impossible* has no place in science. Instead I have made a probabilistic argument: While it is not impossible for the rock to jump up by itself, it is extremely unlikely. Remember, we are dealing with statistical mechanics, so every statement is probabilistic in nature. This is quite different from the physics we learn in school: There is supposed to be only one correct answer, and all others are wrong. When I drop a rock, I know it will fall and not rise. But in reality, it *could* rise—it is just highly improbable, and nobody has yet seen it happen or likely ever will.

Not All Energies Are Created Equal

Impact and friction readily turn kinetic energy into heat, but heat does not easily revert back to kinetic energy. Different types of energy are not always interchangeable. The law of energy conservation tells us that we cannot create or destroy energy, but it does not tell us if a particular type of energy can be converted to some other type. What makes some types of energy more convertible than others?

So far, we have encountered three types of energy: gravitational energy, kinetic energy, and heat (or thermal energy). Each type of energy is associated with certain properties of a system (*system* is physicist lingo for a situation containing objects, energies, and forces). Gravitational energy is completely determined by the height of the object above the ground. Similarly for kinetic energy, the only parameter needed is the speed of the object. However, to completely describe the state associated with thermal energy, we need to know the speeds and locations of all the atoms contained in our system—that is, we would need an astronomical amount of information to fully describe the state of a system that contains thermal energy. Because this is not a realistic proposition, physicists use average values instead. For example, the temperature of a gas is given by the average kinetic energy of the atoms multiplied by a constant. Individually, the atoms in a gas can have different kinetic energies. Since temperature is an average, this tells us little how kinetic energy is distributed among all the atoms of the gas.

Imagine a gas consisting of just five atoms. All five atoms have the same kinetic energy of 40 meV (meV stands for milli-electron volt, a very small energy unit used by physicists). Then the *average* energy of the atoms is 40 meV as well. Now think of a different situation. Another gas also consists of five atoms, but two of them have 5 meV each, one atom has 10 meV, another has 50 meV, and the final atom has 130 meV of kinetic energy. Now, the average kinetic energy is $(5 + 5 + 10 + 50 + 130)/5 = 200/5 = 40$ meV, or the same kinetic energy as our first example. But clearly the situation is not the same. In one case, all the atoms have the same energy, while in the other case, the atoms have very different energies. The difference is the distribution of energy among the atoms. An everyday example of the difference between the average and the distribution of items is household income. The average income in the United States was $50,233 in 2007. But we know that some families scrape by on much less than this, while for others, $50,000 is mere pocket change. The interesting story lies in the distribution, not in the average.

Now, we are ready to measure how convertible energy is. An important property for convertibility is the distribution of energy among all the atoms in a system. All atoms in a rock have about the same height above ground, and therefore, the same gravitational energy. The distribution of gravitational energy in this case is quite simple and can be accurately described by the height of the rock above the ground. But the distribution of thermal energy among all the atoms is much more difficult to describe. Each atom has different energy and vibrates randomly at its own pace. All we know about the rock's thermal energy is its average, given by the temperature, but we know very little about the energy of each atom.

The Curious Case of the Missing Information

When physicists talk about the state of a system, they distinguish between macrostates and microstates. The macrostate of a rock can be described by all the things we know about the rock: its height above the ground, its speed, its temperature, and so forth. The microstate is the exact state of all the parts of the rock, that is, the distribution of speeds and positions of all of its atoms—information we do *not* know. As you can imagine, there are a huge number of microstates compatible with any observed macrostate. Atoms in a rock can wiggle in many different ways, but on average, the rock would still have the same temperature, similar to our example above with the five atoms. Knowing macroscopic parameters, such as temperature, tells us little about the particular microstate of the system.

All this talk of microstates allows us to zero in on a mysterious, yet powerful quantity: entropy. By one definition, entropy is the amount of unknown information about a system or, in other words, the amount of information we would need to fully describe the microstate. Think back to the robber. When I still had my ten dollars, the microstate was very easy to describe: The ten dollars was in my pocket. After the crook stole my money and spent it, the money spread through many hands. The macrostate stayed the same (the amount of money was still ten dollars), but the microstate became more and more unknown (even to the robber). As the robber gave other people part of my money, they in turn spent their money (and so on). The entropy, as it were, of my money increased.

Here is another example: Think of the room of a teenager. The macrostate of his room may be stated as follows: It contains sixty-seven pieces of clothing, twenty-three books, a desk, a chair, a lamp, a laptop, and miscellaneous junk. To describe the microstate, we need to know where all these items are located. There are only a few arrangements (microstates) compatible with a tidy room: clothes in the closet, grouped by long-sleeve shirts, short-sleeve shirts, slacks, jeans, and so on; books on the shelves, sorted alphabetically by author; and so forth. However, there are almost unlimited ways the room could be messy: jeans on top of the computer, stat-mech books on the floor, shirts strewn across the bed. The entropy of a tidy room is much lower (there are fewer possible tidy rooms) than the entropy of a messy room. Now you may ask yourself the same question I ask myself daily: Why is my room always a mess?

This question (not exactly in this form) occupied physicists in the 1800s, and their answer was the second law of thermodynamics. The actual question they addressed is the one we are trying to answer as well: Why are some types of energy more useful than others, specifically, why can some types of energy be converted, while others appear difficult to convert, thus making them useless? Moreover, useful energy eventually turns into useless energy. These unfortunate facts can be understood from our tidy-versus-messy room example. Most of us leave items in random places—places where the items don't belong. Keeping a room tidy requires a lot of work, but without this work, the room inevitably becomes messy. You may wonder, Why does the random placement of items about the room not lead to a tidy room? Why does it always end up messy? Why can't you randomly put your books back in alphabetical order on the bookshelf? This is where our microstates come in: If you leave items lying about in random places, you are more likely to end up with a messy rather than a tidy room because there are many ways (microstates) corresponding to a messy situation and only a few corresponding to a tidy one. In a sense, by leaving items in random places, you randomly pick a microstate from all possible microstates, and because there are more microstates that are messy (and fewer tidy states), chances are high you picked one of the messy states.

Coming back to energy, few microstates are compatible with a certain amount of gravitational energy. This is obvious, as the only relevant parameter is the height of each atom above the ground. Gravitational energy is a low-entropy energy; it is the equivalent of a tidy room. The situation for heat is quite different. Heat is like a messy room; there are so many possibilities for energy distribution among atoms at a particular temperature. Heat is high-entropy energy.

During energy transformation, it is difficult to keep track of the energy among all of the atoms involved. Just like the socks in your room, some energy will be left here, some there—in the form of random atomic motion. Friction and impact are great randomizers of energy. When a rock hits the ground, energy is "lost" to thermal motion of atoms. A tidy situation turns messy. Energy that is completely organized, concentrated, and tidy is a rather artificial, low-probability situation. If we let a system do whatever it "wants," it behaves like an unruly teenager. Energy becomes scattered, dispersed, messy—and unusable.

This tendency of energy to become more and more dispersed and thus unusable is what the second law of thermodynamics is all about. The second law states that in any transformation in a closed system, entropy always increases. The room gets messier, energy disperses, and the money slips through my fingers.

The Second Law

The second law of thermodynamics is one of the most profound, far-reaching, and misunderstood laws of physics (because people do not read the fine print!). Let me restate the second law, but in a more precise fashion: There can be no process whose *only* result is to convert high-entropy (randomly distributed) energy into low-entropy (simply distributed, or concentrated) energy. Moreover, each time we convert one type of energy into another, we always end up overall with higher-entropy energy. In energy conversions, *overall* entropy always increases.

In these statements of the second law, I put certain words in italics—*only* and *overall*. As innocuous as these words may look, they are of great importance in our understanding of the second law and how it relates to life. By ignoring these words, creationists have been able to claim that life and evolution violate the second law of thermodynamics. Not at all!

Let's look at what these words mean, starting with *only*. The second law does not say that a process that converts high-entropy (distributed) energy into low-entropy (concentrated) energy is impossible. If it were, you could not eat ice cream tonight, because a refrigerator is a machine that locally reduces entropy (by cooling things down). But cooling ice cream is not the *only* result of refrigeration. Your fridge also gobbles up electricity (a low-entropy source of energy) and turns most of it into heat (high-entropy energy), which it releases into the kitchen. This is why you cannot cool down your kitchen by leaving the refrigerator door open.

Overall, your refrigerator increases entropy by a large amount even though it locally decreases entropy. You can locally reduce entropy, but you need to do a lot of work and consume a lot of low-entropy energy in the process (think of tidying your room—it is not impossible, but you end up sweating and cursing—and increasing the entropy of your surroundings by burning low-entropy food energy and turning into high-entropy heat and waste). The same applies to how life works: Life uses a low-entropy source of energy (food or sunlight) and locally decreases entropy (creating order by growing) at the cost of creating a lot of high-entropy "waste energy" (heat and chemical waste). The creationist claim that the second law does not allow for life or evolution does not hold water.

Entropy and the second law of thermodynamics are among the most important concepts of physics and among the most misunderstood. Part of the problem is there are several definitions of entropy and none are straightforward. Often, entropy is simply equated with disorder. This is not a good description unless we clearly understand what is meant by disorder. Again, creationists have exploited this confusion, claiming increases in entropy are incompatible with life. Their argument is that because entropy is disorder and life is order, the second law proves that life could not have emerged spontaneously.

Equating entropy with disorder is convenient, but it is not a precise definition by any means. Scientists run into this problem frequently. When casting (usually mathematical) definitions into everyday language, the details get lost in translation. Entropy is not the same as the disorder we think of every day (our tidy-versus-messy room example was only an analogy to explain the concept of microstates). Instead, entropy measures the degree to which energy is spread out. Sometimes an orderly-appearing system may have energy that is more dispersed than that of a "disordered" system. Such a system, although seemingly more ordered, would have higher entropy.

A surprisingly simple example illustrating this difference between entropy and the everyday concept of disorder is a collection of hard spheres (think of marbles). As we put more and more marbles into a container, the marbles reach a critical density (number of marbles per volume) at which the highest entropy state (i.e., the one we would expect to be most disordered) is the state in which the marbles are perfectly stacked in an ordered pattern. How is this is possible? Marbles can be filled into a container in two fundamentally different ways: We can just pour them into a jar and let them fall where they fall—or we can carefully stack them into an ordered array. Just pouring them leads to a situation where marbles are touching, but are otherwise in random positions. Such an arrangement is called *random stacking* by physicists. When marbles are randomly stacked, some marbles will become completely wedged and will not be able to move. Random stacking reduces freedom of motion, leading to a more narrow energy distribution and thus lower entropy. Nicely stacked marbles, on the other hand, have on average a little bit more wiggle room, and therefore higher entropy. Marbles are a good illustration of a simple system where higher entropy means more order, not less. Simply equating entropy with disorder can be misleading. In biological cells, there are many ordered structures that form spontaneously while increasing entropy. These include assemblies of proteins, cell membrane structures, and fibers. In these cases, the entropy is increased by exporting the disorder to the surrounding water molecules. More about how this works in Chapter 4.

Once again, we find that the second law does not preclude the emergence and continued presence of life. Entropy can be reduced locally if it is increased globally (in our refrigerator, for example), and sometimes, an increase in entropy (hard spheres, for example) leads to more order. Life takes advantage of both of these apparent loopholes of the second law.

Free Energy

Life reduces entropy locally while increasing it globally. The concept of entropy describes this situation reasonably well, but it has one drawback. To square life's processes with the second law of thermodynamics, we need to analyze the entropy of an organism *and* the entropy of its surroundings (because the surroundings ensure that the second law is not violated). However, details about the environment may not be well known, and all we really want to know is whether heat (and entropy) can flow to the environment or not. For this we only need to know the temperature of the surroundings. This fact led scientists to establish a new way to analyze thermodynamic systems that are immersed in constant-temperature environments (for example, a living organism or a cell).

How can the temperature of a system be kept constant? If you are chemist and your system is a test tube, you can place the test tube into a large bath maintained at constant temperature (for example, a bucket of ice and water at 0 degrees Celsius). If the system begins to heat up from an exothermic (heat-releasing) reaction, the heat will quickly flow into the surrounding bath until the system's temperature equals that of the bath. If the system cools down (an endothermic reaction), it will draw heat from the surrounding bath. If the bath is large enough, the heat flow from the system will not change the temperature of the bath. Thus the bath serves as an energy reservoir and a temperature control.

Living organisms also act as heat baths. Our cells are immersed in a large 37 degree Celsius temperature bath—our bodies. All chemical reactions in our cells happen at this constant temperature. If we want to apply the second law to this situation, it becomes tricky. The second law states that *overall* entropy increases. Therefore, if I look at a reaction happening in one of my kidney cells, I would need to know not only how the entropy of the participant molecules in the reaction changes, but also how entropy of the surrounding tissue changes. I would need to analyze the change of entropy in legions of complicated cells just to determine the outcome of one tiny reaction.

There is an easier way. The exchange of energy and entropy with the surrounding heat bath can be represented simply by the temperature, as long as we know the temperature is kept constant by the bath. To do this, we use the concept of *free energy.* Free energy F is the total energy E minus the product of temperature T and the entropy S of the system (or $F = E - TS$). Because entropy represents how much energy has become dispersed and useless, free energy represents that part of the energy that is still "concentrated" and useful (because we are subtracting the useless part, TS).

To analyze a system within the constant-temperature bath of our own bodies, we need to know how the system's free energy changes. In the language of free energy, the second law is restated this way: At constant temperature, a system tends to decrease its free energy, until, at equilibrium, free energy has reached a minimum. The second law tells us that useful energy will become degraded, and eventually we will only be left with dispersed, unusable energy.

Because free energy includes both energy and entropy, the concept of free energy explains many of the paradoxical situations we discussed above. For example, in the random-stacking example, there is not much difference between the energy of marbles stacked randomly or orderly, but the entropy is higher for the ordered stacking. Therefore, the free energy, given by energy *minus* entropy times temperature, is *lower* for the ordered marbles, and because the system tends to minimize its free energy, it will tend to an ordered state at equilibrium.

We can also think of examples where both the entropy and the free energy of the system are lowered. What is so special about this? When we lower entropy, the free energy tends to go *up*, because we subtract the entropy term from the total energy. For example, if the total energy of a molecule is 5 eV, and entropy times temperature is 2 eV, then the free energy is $F = 5$ eV $- 2$ eV $= 3$ eV. Now, if the entropy term decreases to 1 eV (for example, by cooling the system), the free energy would *increase* to 5 eV $-$ 1 eV $=$ 4 eV.

Can both free energy and entropy decrease at the same time? Keeping the temperature constant, this is only possible if the total energy of the system, *E*, also decreases, and decreases more than the entropy term. For example, if total energy decreases from 5 eV to 3 eV, while the entropy term decreases, as before, from 2 eV to 1 eV, the new free energy would be 3 eV $-$ 1 eV $=$ 2 eV. This is less than the original 3 eV, despite the fact that entropy is decreased.

Why doesn't this example violate the second law? Isn't entropy supposed to increase? When the total energy of the system changed from 5 eV to 3 eV, the "missing" energy had to go somewhere (remember energy conservation). Energy passed into the environment, heated it up, and increased the environment's entropy. As before, *overall* (system + environment) entropy increased, even though, by itself, the entropy of our small system decreased. You can see how the language of free energy makes it easier to think about such a scenario: As long as the free energy of a system decreases, we are obeying the second law.

Does nature provide examples of spontaneous decreases in entropy? All the time! Take the creation of a snowflake. Compared with a liquid drop of water, a snowflake has much lower entropy (it is also much more ordered). Yet, snowflakes form spontaneously under the right conditions. This is because the energy (*E* in our formula) of a snowflake is much lower than the energy of a water droplet. Once the temperature has fallen below the freezing point of water, the entropy reduction is overwhelmed by this reduction in energy, and the free energy is reduced when water freezes into beautiful, ordered crystals. Thus, at low temperature, a snowflake has lower free energy than a water droplet. At high temperature, however, the entropy term wins again (it is multiplied by the now higher temperature) over the energy (*E*) term, and water turns liquid. At higher temperature, liquid water has lower free energy than frozen water. This is shown in Figure 3.1.

We can also think about the competition between liquid and frozen water in the language of the molecular storm. At low temperatures, the forces between atoms are stronger than the shaking of the molecular storm and draw atoms together to form structures; at high temperatures, the forces between atoms are no match for the more violent molecular storm, and snowflakes melt. The concept of free energy captures the tug-of-war between deterministic forces (chemical bonds) and the molecular storm—or in other words, between necessity and chance, in one elegant formula, $F = E - TS$.

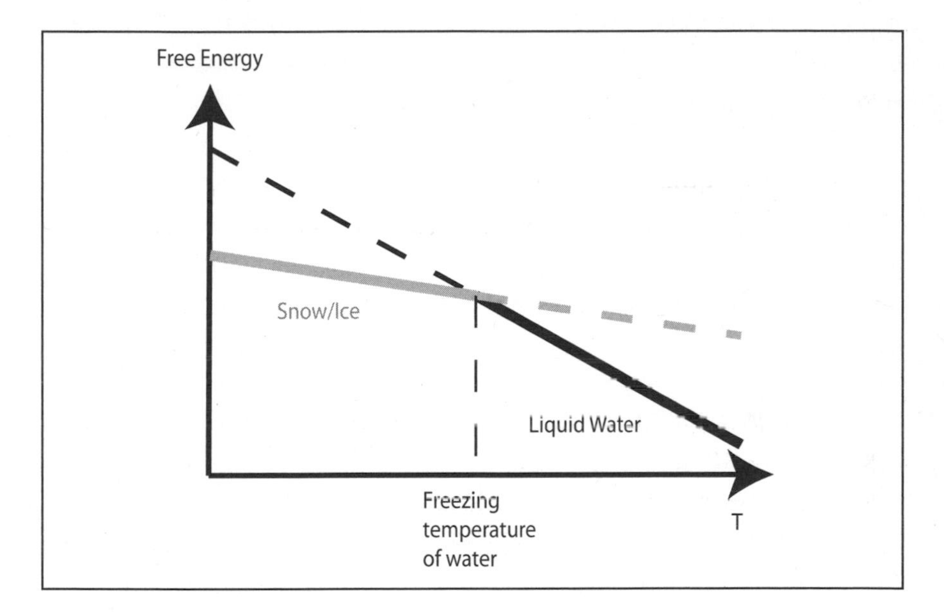

FIGURE 3.1. Free energy of liquid water (black) and ice or snow (gray) plotted against temperature. Because liquid water has higher entropy, it is represented as a steeper line in the diagram. At low temperatures, ice has lower free energy than liquid water, and water freezes. At high temperature, liquid water has lower free energy (because of the higher entropy), and ice melts.

Driving the Molecular Storm

According to the second law, free energy will eventually be degraded and reach a minimum. If this tendency holds for all natural processes, then the universe must have started out with an abundance of free energy at the time of the big bang. This is clearly the greatest gift of the universe: Without this gigantic amount of free energy, the universe would be a cold, uniform, and very boring place. How does the universe share its bounty? The big bang started out as pure energy and very little entropy (a single point cannot have much entropy). Shortly after the big bang, energy congealed into energetic particles that whizzed around near the speed of light. The particles collided, annihilated, or appeared out of the expanding fireball, creating chaos where formerly there was simple order. In this expanding fireball, the newborn universe started to create entropy (as well as degrade some of its free energy). As free energy was degraded, new structures were brought into existence—according to Einstein's equivalence of mass and energy ($E = mc^2$), featureless energy became quarks, electrons, muons, neutrinos, photons, and all their tiny brethren.

After three hundred thousand years, the universe had cooled enough to form the first atoms: Quarks bound together in threesomes called protons and neutrons. Protons and neutrons clustered into nuclei, which captured electrons. Again, the entropy of the universe increased with the release of atomic binding energy (which turned into heat), and something new was created in the process.

The universe continued to cool and became less and less dense. Empty, cold, and no longer able to form nuclei of atomic mass higher than hydrogen and helium, the universe filled to more than 99.9 percent with just these two elements. There is a joke among astronomers that the periodic table really only needs those two entries. All the other elements are so rare, we may as well neglect them (of course, there would be no astronomers if that were truly the case).

Filling the universe with hydrogen and helium was like filling it with fuel. When light nuclei combine in a process called nuclear fusion to form heavier nuclei, they release large amounts of energy (this is what happens in a hydrogen bomb). To make heavier nuclei, two lighter nuclei must collide at very high speed. High speed implies high temperature; as the difficulty of building a nuclear fusion reactor demonstrates, the needed temperatures and densities are difficult to achieve.

This is where gravity lent nature a helping hand. In the early universe, some regions (just by chance) had slightly higher densities than neighboring regions. These denser regions gravitationally attracted more atoms, gained more mass, and continued to attract more atoms. A once-almost-even distribution of atoms became increasingly clumpy. Atoms clustered together in giant nebulae, separated by giant voids. The nebulae began to collapse under their own weight and, with their continued collapse, became hotter and hotter. Finally, densities within the nebulae became so great and the collisions between nuclei so fast that nuclear fusion began to create heavier nuclei. Hydrogen and helium were cooked into heavier elements, and stars were born.

Stars are the furnaces that "burn" the overabundance of fuel in our universe. Deep within stars, nuclear fusion creates heavier and heavier nuclei, all the way up to iron (even heavier elements are created in large stellar explosions, called supernovae). In fusion, energy is released and streams outward in the form of radiation. Our sun bathes the Earth's atmosphere in electromagnetic radiation, or light, which is absorbed by molecules and atoms. When molecules absorb this light energy, it is ultimately converted to kinetic energy, making the molecules shake, rotate, or move faster. As these faster molecules collide with slower, less energetic molecules, the faster molecules give up some of their energy in the collision. Soon, this gift of energy from the sun is distributed among many molecules, heating up our atmosphere. The molecular storm, and the abundance of free energy on our planet, come forth from the universe, carried to us by our sun.

Open Systems

This short history of the universe has shown that the degradation of free energy is not all bad. As free energy is dissipated, and the entropy of the universe increased, new structures are born, from quarks to nuclei to atoms to . . . life. The universe continues to share its free energy with abandon. The continual flux of energy is a fact of life—a fact that keeps living systems out of thermodynamic equilibrium. Equilibrium is the state in which all available free energy has been degraded and no usable energy remains. Equilibrium means death. Living beings must avoid equilibrium. As long as we are alive, energy continues to flow through us. In thermodynamics, systems through which energy and matter flow from and to the environment are called *open systems*.

Recognizing that living organisms are open systems is an important step toward our understanding of life—but it is not enough. A small volume of gas within a larger one is also an open system; new molecules enter this volume all the time, while others leave. If left alone, the volume will tend to a state in which, on average, the same amount of molecules enter and leave at any point in time. Moreover, the molecules are not transformed. The molecules that enter are the same as the ones the leave. Consequently, there is no degradation of free energy. Although such a volume is open, it is at equilibrium.

Living systems are different: What enters is not the same as what leaves the system. Living beings gobble up low-entropy energy, degrade the energy, and expel high-entropy energy into the environment. We call such systems *dissipative systems*, because they continuously dissipate free energy into high-entropy energy.

Thus living organisms are *open, dissipative* systems. But there is still more. Nature is full of open, dissipative systems that we would not consider alive. A hurricane is an example. It takes a low-entropy source of energy (the large temperature difference between the ocean and the upper atmosphere) and continuously dissipates this energy in a display of awesome power. It dissipates the energy by moving the heat of the ocean to the cool upper atmosphere by convection (air-mass movement due to differences in temperature). The motion of huge masses of air, coupled with forces originating from the Earth's rotation, soon organize the moving air into a giant rotating storm. As long as there is a supply of warm ocean air, the storm continues to rage, dissipating the heat energy of the ocean as it sweeps across the water. What is most striking about the hurricane is its structure: the rotating swirls of clouds, the eye of the storm in the middle. Many open, dissipative systems show spontaneous emergence of structure, in seeming violation of the second law. But we already know there is no contradiction here. The hurricane increases entropy *overall* far more than it locally decreases it.

The study of far-from-equilibrium systems, and their spontaneously created "dissipative structures," has led many scientists to speculate that living systems are similarly built. But that would be misleading. A close look at life at the microscopic level shows that it is a tightly controlled dance of sophisticated molecules, designed by evolution. It is not a spontaneous, wasteful system like a hurricane. Life is a highly efficient process. Efficiency is best achieved when we do not stray too far from equilibrium, because large movements cause friction and, consequently, rapid degradation of low-entropy energy. Life chooses the middle road: By staying away from equilibrium, we stay alive. By staying close to equilibrium, we increase efficiency.

Life is a near-equilibrium, tightly controlled, open, dissipative, complex system. Such a system can only work if its parts are "designed" (by evolution) to push thermodynamics to its limits. Life does not exist despite the second law of thermodynamics; instead, life has evolved to take full advantage of the second law wherever it can. But how can it do this? Life's engines operate at the nanometer scale, the tiny scale of molecules. But what is so special about this scale that chaos can become structure, and noise can become directed motion?

4
On a Very Small Scale

There is no excellent beauty that hath not
some strangeness in the proportion.

—FRANCIS BACON

A biological system can be exceedingly small. Many of the cells are very tiny, but they are very active; they manufacture various substances; they walk around; they wiggle; and they do all kinds of marvelous things—all on a very small scale.

—RICHARD FEYNMAN, "THERE'S PLENTY OF ROOM AT THE BOTTOM," LECTURE AT AMERICAN PHYSICAL SOCIETY MEETING, 1959

I had been impressed by the fact that biological systems were based on molecular machines and that we were learning to design and build these sorts of things.

—K. ERIC DREXLER

Richard Feynman quotation by permission of California Institute of Technology, *Engineering and Science Magazine*. K. Eric Drexler quote courtesy of K. Eric Drexler.

IN MARCH 2011, I ATTENDED THE BIOPHYSICAL SOCIETY meeting to learn what the veterans of this field have been up to and where I should take my own research. Biophysics deals with the physical underpinnings of living systems and the use of physical methods to explore life. When biophysics was founded in the 1800s by researchers such as Helmholtz, it dealt with the parts of living organisms one could easily handle and see. It was a macroscopic science. Today, when you attend the largest meeting of biophysics on the planet (sixty-five hundred participants, over seven hundred posters, every day, for four days), absolutely everything deals with structures only found at the nanoscale. Biophysics today is nanophysics.

The biomolecular world is filled with exquisite structure and a mysterious drive for change and motion. In several talks, people presented *motility assays*, a fancy term for attaching protein molecules called myosins to a surface and then seeding fibrous proteins, called actins, on top of the myosins. All these molecules are so tiny that they cannot be easily seen in an optical microscope. To make them visible, the researchers attach little molecular flashlights, fluorophores, to them, which turn the actin filaments into molecular fireflies. Immerse the myosin and actin filaments in a liquid buffer solution, and not much happens, but add an energy storage molecule, called ATP, and all over the surface, actin comes "alive." Like little nanometer-scale worms, the actin filaments start moving in almost straight paths. Sometimes they hit an invisible obstacle and curve in a new direction. As long as enough ATP is provided, they just keep going and going. What moves them? According to the researchers, it's the myosin molecules. They act like molecular motors, pushing actin forward with their two molecular "hands" and passing each actin filament from one myosin molecule to the next, like a rock star crowd-surfing.

In other talks, researchers explained how cells change their shapes by polymerizing filaments such as actin or the sturdier microtubules. Filaments are made of small units, which spontaneously form by assembling themselves (polymerization) or by being actively assembled and disassembled by molecular machines. As the filaments grow, they push on the cell surface, creating protrusions. This is how cells move. Yet another set of talks dealt with the intricate structures of cell membranes, the thin shell that surrounds each cell and separates the inside of the cell from the outside environment. Riddled with specialized pores, the membrane only admits desired molecules into the cell, while undesired molecules are kicked out. Floating on the membrane are the cell's "TV antennas": cell receptors waiting for a chemical signal from the outside world. Once a signal arrives, it is transmitted through the membrane, setting up a cascade of activity that may lead to cell motion, cell division, the secretion of a compound, or cell suicide. This is the nanoscale world of our cells.

Life *must* begin at the nanoscale. This is where complexity beyond simple atoms begins to emerge and where energy readily transforms from one form to another. It is here where chance and necessity meet. Below the nanoscale, we find only chaos; above this scale, only rigid necessity.

Nano

How do you tell a bunch of sixteen-year-olds how small a nanometer is? I was standing in front of eighty high school juniors from the Macomb County Math and Science Center, trying to explain my research. I needed to get them to imagine the unimaginable. "A nanometer is to the size of a human, as the size of a human is to ten times the distance from the earth to the moon," I began, but quickly realized that the distance from the Earth to the Moon is not something many of us have experienced firsthand. Little pearls of sweat started forming on my forehead. I tried again: "A nanometer is so small, you would need to slice the width of a human hair one hundred thousand times to reach a nanometer." Better. But how about translating distance into time? "If I would shrink you to one nanometer in height, you could walk about twenty-five hundred nanometers in one hour. At this speed, it would take you eighty-two *years* (!) to walk the length of a full-sized human, from his toes to the top of his head." Gasps. They started to realize that a nanometer is not just small—it is so small that the nano-realm is utterly removed from anything we could ever hope to experience. Yet, we can measure this stuff.

Fifty years before my little lecture to future scientists, Richard Feynman, the famous Nobel Prize–winning physicist, gave a groundbreaking lecture to fellow physicists at the 1959 American Physical Society meeting. In typical Feynman fashion, the lecture was simply titled "There's Plenty of Room at the Bottom." The "bottom" was the microscopic scale, from micrometers (one thousand nanometers) down to atoms at just a few tenths of a nanometer in diameter. Feynman's point was that no law of physics should keep us from creating machines that are just a few nanometers large. It's simply a question of engineering.

It took a while for Feynman's vision to become reality, but by the late 1980s, nanotechnology was starting to take off. With the invention of tools to image objects only a few nanometers in size and to measure and manipulate them in various ways, it was now possible to compress data to nanoscale bumps or to build ever-more-complex nanostructures. These advances led to a kind of frenzy of wild predictions in the 1990s and into the 2000s, some overhyped (nanotechnology as the savior for all our energy, medical, and environmental problems) and some doomsday (gray goo of nanorobots eating everything in sight—as in the remake of *The Day the Earth Stood Still* or in Michael Crichton's book *Prey*).

At least in the media, the nanotechnology craze has died down a bit. The early promises of nanorobots (or nanobots) cleaning plaque out of our trans-fat-challenged arteries have not materialized as fast as expected. The dangers of nanotechnology are there (small nanoscale fibers can possibly cause cancer—think asbestos), but the gray-goo idea seems highly overdrawn. The media have moved on. But in science, nanotechnology and nanoscience are alive and well. Scores of physicists, engineers, chemists, and medical researchers are engaged in nanotechnology research, from nanobatteries to nanomedicine.

Personally, I prefer to speak of nanoscience rather than nanotechnology. Nanotechnology is the next step, after the science has been worked out. What is nanoscience? In short, it is the production, measurement, and understanding of systems where at least one spatial dimension is in the nanometer range. Sometimes, this broad definition has led to trouble, as many old areas of research, such as thin-film technology and some branches of chemistry, suddenly became nanotechnology research, only because they dealt with things smaller than one micrometer. Thus there is a joke about nanotechnology—that it is simply a ruse to get money out of funding agencies. While there was some truth to this charge—at least in the early days—there is something genuinely special about the nanoscale: Systems, once shrunk down to this "magic" scale, exhibit new and rather unexpected properties.

Feynman's talk at the 1959 American Physical Society meeting is often credited as having jump-started the nanoscience revolution. The truth, however, is a bit more complicated. When he gave his famous talk, the assembled listeners did not take the topic very seriously. One attendee of the meeting recalls: "The general reaction was amusement. Most of the audience thought he was trying to be funny . . . It simply took everybody completely by surprise."* Feynman's talk was rediscovered twenty-five years later, when many of his predictions had come to pass. By then, technology had caught up with many of Feynman's visionary ideas, and they were finally taken seriously.

Feynman took his inspiration from living systems, as have many recent nanotechnology visionaries. He was taken by the way information was written "on a very small scale" in biological cells, and how cells used the information to "manufacture substances," "walk around," and "do all kinds of marvelous things."** Indeed, if today's nanotechnologists are dreaming of building nanosize machines, they have to accept that nature beat them to the punch by a mere three billion years! Living cells are teeming with molecules that perform amazing feats at a nanoscale with almost uncanny precision.

Feynman envisioned that nanoscale miniaturization would allow us to store whole books on the head of a pin, build tiny motors, move single atoms around, or build powerful pocket-size computers. All of these things have come to pass. Some of his other predictions are still not feasible, such as the creation of a nanosize surgeon, an idea we now know as the nanobot. A nanobot would be a tiny device that could be swallowed like a pill or injected into the bloodstream. The device would perform nanosurgery, such as cleaning plaque from arteries or performing search-and-destroy missions on cancer cells. However, in a field called targeted drug delivery, or nanomedicine, researchers have already made remarkable progress

---

* Christopher Toumey, "Reading Feynman into Nanotechnology: A Text for a New Science," *Techné* 12, no. 3 (fall 2008): 133–168. Tourney ascribes the quote to Paul Shlichta, then of the Jet Propulsion Laboratory, California Institute of Technology, Pasadena.

** Richard Feynman, "There's Plenty of Room at the Bottom," *Caltech Engineering and Science* 23 (February 1960): 22–36, available at www.zyvex.com/nanotech/feynman.html.

in creating nanostructures which will specifically seek targeted cells (for example, cancer cells) and then deliver their payload (deadly drugs to a cancer cell, or a piece of DNA to repair gene damage) only to the targeted cells. Another one of Feynman's predictions, which he took from a science fiction story by Robert Heinlein, was the possibility of building a machine that could construct a smaller version of itself. The smaller version, in turn, could build an even smaller version. Like a set of Russian nesting dolls, the machines would build smaller and smaller versions of themselves, all the way down to the last, nanometer-size machine.

Similar ideas of tiny machines made of molecular gears propelled an MIT engineer, K. Eric Drexler (b. 1955), to write a now-famous book about the possibility of molecular machines. When he read Feynman's talk in the late 1970s, Drexler had already thought about the possibility of making machines out of molecules, and reading Feynman gave him an additional impetus. In 1986, he published the founding work of modern nanotechnology: *Engines of Creation: The Coming Era of Nanotechnology*. Much of Drexler's ideas have so far remained science fiction, although there has been substantial progress in some areas. For example, moving single atoms and molecules around one by one and making structures such as circles or triangles made of atoms on a surface is, if not routine, certainly an achievable feat with the right tools.

The right tools were being invented about the same time Drexler wrote his books. The most iconic of these tools, scanning probe microscopes, had a great impact on the burgeoning science of the nanoscale and an equally great impact on my own life.

Touching Atoms

When you defend your Ph.D. at the University of Oxford, you'd better know your stuff. Examiners are brought in from other universities, preferably from outside the country. They don't know you, but are there to ensure you're no slacker. When my friend Steve got his Ph.D. at Oxford from his work with atomic force microscopes (AFMs), his examiners were two of the best-known AFM experts in the world: Christoph Gerber, who together with Nobel Prize–winner Gerd Binnig, had invented and built the world's first AFM in 1986; and Ernst Mayr, who runs one of the largest and most successful AFM groups in the world. Gerber, who works for IBM, is a technical genius. When he visited Oxford in 2000 to be an examiner for Steve's Ph.D. defense, he told us about a new AFM spinoff: an artificial nose. This nose works by coating tiny cantilevers (micrometer-long beams of silicon) with different substances that absorb airborne chemicals. When the chemicals are absorbed, the substance on the cantilever expands, and the cantilever bends. Using an array of differently coated cantilevers, each sensitive to different types of chemicals, this mechanical nose can be trained to distinguish many different smells. Gerber, a connoisseur of Scottish whiskeys, was excited to find that his mechanical nose had no problem distinguishing a Craigellachie from a Laphroaig, but he was even more surprised when the nose told him that one of his whiskeys had a hint of cherry. He called up the distillery in Scotland, and, indeed, the whiskey had been aged in cherry wood casks.

The AFM is the most common member of a group of instruments called scanning probe microscopes (SPMs). SPMs have revolutionized the measurement and manipulation of matter at the nanoscale more than any other instrument invented since Feynman's famous talk. The first SPM was the scanning tunneling microscope (STM), invented in 1982 by Gerd Binnig, Christoph Gerber, and Heinrich Rohrer at the IBM laboratory in Zurich, Switzerland. The capability of this new instrument to image single atoms and measure their electronic properties was so astounding that Binnig and Rohrer received the 1986 Nobel Prize for their invention.

The working principle behind SPMs is surprisingly simple. Instead of detecting light from an object, as we would do in conventional microscopy, SPMs "feel" the surface by means of a sharp probe. The way the SPM feels the surface depends on the nature of the probe: In a STM, the microscope feels a "tunneling current" between the probe (a sharp metallic needle) and a conducting surface. Tunneling, a purely quantum-mechanical effect, is the ability of an electron to traverse an energy barrier it should not be able to cross, according to classical physics. In STM, this energy barrier is posed by the gap between the probe and the surface it is imaging. The tunneling current measured by STM depends on the width of the barrier (the distance between the probe and the surface) and the amount of electrons in the sample. Measuring the tunneling current across a sample surface provides a map of the changing heights on the surface, or changing electron density, or both. Tunneling is so sensitive to tunnel barrier width that each time the probe is moved away from the surface by only a tenth of a nanometer, the tunnel current decreases by a factor of ten! This extreme sensitivity allows the STM to obtain images of single atoms.

The images obtained with an STM are a convolution of changing heights on a surface and the changing electron densities (which vary around atoms). However, that's not all. Forces between atoms on the tip and surface also play an important role. This was becoming clear soon after the invention of STM, when eager researchers around the world built their own STMs to look at a plethora of different surfaces. On some surfaces, the images obtained did not square with theoretical predictions taking only tunneling barrier widths and electron densities into account. One group of researchers, for example, noted that on some surfaces, the forces were so high that the distance calibration of the instrument was thrown off. This problem was addressed with the invention of a second, and now the most popular, scanning probe microscope, the AFM.

The (E)squires of Oxford

The scientist who inspired the creation of the AFM was John Pethica, of the University of Oxford, who fifteen years later became my postdoc advisor. Gerber called him the godfather of the AFM. (Which makes me, by scientific genealogy, the nephew of the godfather of the AFM, which probably is not much to brag about.) At Oxford, under the benevolent, inspirational, and deliberately hands-off guidance of John Pethica, my friends Steve Jeffery, Ralph Grimble, Ahmet Oral, Özgür Özer, Chandra Ramanujan, and I learned how to be innovative scientists. We built and used STMs and AFMs (actually combining the two) and measured the nanomechanics of atoms on crystal surfaces and molecules in liquids. It was a good time.

John is an easygoing person, but he is an exacting scientist. When I visited Oxford for my interview, an envelope with instructions was waiting for me at the hotel. John had written "Dr. Peter Hoffmann, Esq." on the envelope. I didn't know that I was an Esquire, but it showed John's respect for everybody, even a lowly postdoc looking for a job. When my wife and I finally arrived a few months later, it was the beginning of summer. John typically disappeared for extended periods, only to reappear with a bagful of new ideas. After John's return from his mysterious summer travels, I got into one of the typical—as I soon realized—conversations with him. These conversations always involved new ideas, connections, and recent publications. Listening to John, I would often be reduced to nodding and saying, "aha, yeah, mmmh," only to scramble back to the office to look up the papers he was talking about.

John is particularly interested in how we make the transition from the noisy but reversible world of the atom, to the more ordered but typically irreversible world of macroscopic objects, a passion I inherited. In one experiment, we measured the loss of energy an oscillating AFM tip experiences when it interacts with randomly oscillating atoms on a surface, in a process we called *atomic-scale energy dissipation*. For small numbers of atoms involved, the motions of the atoms seemed structurally reversible. Each time we lowered the tip, we measured the same force curve. Yet when we pushed harder, we passed the threshold to permanent rearrangement of atoms. When too many atoms were involved, there were too many possible configurations, and the system did not find its way back to its original arrangement. At this point, our experiment had passed from the microscopic to the macroscopic.

The interaction of atoms and the myriad possibilities of atomic arrangements in modestly large systems is also of great importance in molecular biology, as we will see. So is the transition from noise to order. Small nanoscale systems, such as the molecules in living cells, are subject to influences and laws that are quite different from what we encounter in our familiar, macroscopic world.

The Incredible Strangeness of Small Things

When Feynman pushed for the creation of atomic-scale technology in his talk, he also pointed out unique challenges that come with building machines out of just a few atoms. Indeed, since nanoscience has become a serious science, the most interesting feature has been that tiny things play by different physical rules. This sometimes creates problems and, other times, opportunities, but it is a source of continuing fascination to nanotechnologists.

When systems are shrunk to the nanoscale, effects that play little or no role at the macroscale suddenly become important. In his famous lecture, Feynman mentioned a few of these: problems of lubrication—because he surmised that the effective viscosity of any lubricant is much higher at a small scale—and the observation that at small scales, things are more likely to stick together. These problems are due to two more general effects that are common at the nanoscale: the graininess of matter and the problem of interfaces.

Matter is grainy: it is made of atoms, molecules, and larger grains, such as small crystals. On a large scale, this graininess of matter averages out. Following the laws of statistical mechanics and *continuum physics* (in continuum physics, we ignore the fact that matter is made of particles, but treat it as a smooth continuum), we can determine such average quantities as Young's modulus (which determines how springy a solid material is) or viscosity (how resistant to flow a liquid is). But once we reduce the size of a system to just a few, ten, or one hundred molecules, these averaged quantities become meaningless, and measurements of mechanical or electrical properties show jumps, rather than smooth, averaged-out changes. In my lab, we measure the graininess of simple liquids, such as water. When I measure the mechanical properties of water, it doesn't matter if I take a bucketful of water, or a few cubic centimeters from a syringe. A cubic centimeter of water contains 3×10^{22} water molecules (that's a 3 with 22 zeros behind it). Adding or subtracting a few million or billion molecules from such a giant number of molecules is not going to change the average properties of the liquid. In our lab, however, we can measure the mechanical properties of much smaller numbers of water molecules. We can squeeze water between an AFM tip and a surface until the water layer between the tip and surface is a single molecule thick (however, since the tip has an area of about 50 nanometers by 50 nanometers, the single molecule layer under the tip contains about 25,000 molecules). Now, adding or subtracting single layers of molecules makes a huge difference to the mechanical properties of the liquid. As a result, when we push from a layer 6 molecules thick to one 5 molecules thick, and then from 5 to 4, and so on, the stiffness and the apparent viscosity of the layers alternate between high and low values.

These experiments also illustrate the other nanoscale problem Feynman was talking about. When we make something smaller, its surface-to-volume ratio increases. A golf ball has a greater surface-to-volume ratio than does a bowling ball. Shrunk to the nanoscale, this ratio would be more extreme. As volumes become small, surfaces start to dominate and the forces that are important at the macroscale become irrelevant at the nanoscale, and vice versa. At the macroscale, forces associated with mass, such as gravity and inertia, dominate. Surface forces, such as stickiness, are usually unimportant, unless specifically engineered, as in a glue. For example, in a baseball game, inertia (when the bat hits the ball) and gravity (when the ball comes back down) dominate. But, typically, the baseball does not stick to the bat. In a game of nanobaseball, however, inertia would be unimportant, as the ball would weigh next to nothing. Ditto gravity. But the relatively large surface area compared with the tiny bulk of the nanobaseball would make it difficult to get the nanobaseball off the nanobat. This is an example of one peculiar property of nanoscale systems: profound changes in behavior depending on the size of the system.

Of course, not everything at the nanoscale sticks together. Otherwise, we'd be in trouble. Our cellular components have to be able to stick and separate when needed. This is achieved through a careful balance of forces between molecules and the surrounding salty water. In a vacuum, in the absence of water, most surfaces simply tend to stick.

At the nanoscale, there are some peculiarities that Feynman did not mention: quantum-mechanical effects, the importance of thermal noise and entropy, cooperative dynamics, large ranges of relevant time scales, and the convergence of energy scales. A daunting list of strange properties. We will discuss these peculiarities through different examples in the remaining chapter.

Quantum-Mechanical Effects

Most books on nanotechnology will focus on the strange quantum-mechanical effects we encounter at the atomic scale. I alluded to some of these in the discussion of tunneling. Indeed, a large part of nanotechnology relies on new quantum-mechanical effects. However, for molecular biology, these are almost irrelevant. Essentially all of molecular biology can be explained using classical physics (except bonding between atoms, which requires quantum mechanics). Many of the more interesting quantum-mechanical effects in nanosystems are completely destroyed by thermal motion. Therefore, much research on quantum computing, spintronics, or other fancy new quantum electronics is done at low temperatures—much too low for any living system.

Thermal Noise

By contrast, thermal motion (or, what physicists like to call thermal noise), already discussed in the previous chapter, is of great importance in biology and most of room-temperature nanotechnology. So are questions of size dependence, cooperative dynamics and time scales (which go hand-in-hand), and the convergence of energy scales. To understand what these things are, and how they influence systems at the nanoscale, let us consider the important question: How can we make stuff at the nanoscale?

Assemble Thyself!

Feynman believed machines and structures at the nanoscale could be assembled by some kind of demagnifying technology that could turn a macroscopic template into a small pattern using electron or ion beams (now a reality in devices called electron-beam writers or focused-ion beams) or by building machines that make smaller copies of themselves ad infinitum until the nanoscale is reached (not yet a reality). Both of these approaches could be described as top-down: They start with a large template or machine and then miniaturize down to the nanoscale.

This top-down approach is still used in building electronic devices, such as computer memory or processors. Life, however, works differently: Plants, animals, and all other living organisms are built from the bottom up. We are all assembled atom by atom, molecule by molecule.

A common fallacy that people hear over and over is that our DNA contains all the information needed to make a human being. Nonsense! The amount of information contained in our DNA is staggering, but it is not nearly enough to specify each molecule's or cell's location, or even the shape of an organ. Rather than being a blueprint (as DNA is often mistakenly called), DNA is more like a cooking recipe. When I make a cake, I don't have to specify where each starch or sugar molecule goes. I just follow the instructions, and the molecules go where they are supposed to. Much of the information to make a cake or a human being is contained in the laws of physics and chemistry. Molecules "know" how to put themselves together.

This self-assembly of molecules is ubiquitous throughout nature and is a major research area in nanoscience. If we could coax molecules to arrange themselves into any structure we want—much like what we see in living organisms—we could make new devices incredibly cheaply, without million-dollar ion or electron-beam writers. We could simply put everything in a pot and stir. But it's obviously not that simple.

One of my recent graduate students, Venkatesh Subba-Rao, likes to tell the following anecdote: He once listened to a talk by a visiting scientist about structures in liquid crystals. When Venkatesh asked him the reason why these structures arise, the scientist answered, "They minimize free energy!" Since then, if I ask my students why this or that is happening in their experiments, and they don't know, this is their stock answer. Most of the time it's correct. Everything that happens in nature minimizes free energy. But this answer does not tell us much. It is almost the scientific equivalent of "because God made it so." A more useful answer would include the types of energies that make up the free energy of a system. An even more useful answer would identify how molecules move around and how energy is transformed as the system minimizes its free energy. Remember, free energy is the difference between the *total* energy of the system and *unusable* thermal energy, leaving the usable, free part of the energy. Unusable energy is the product of temperature and entropy. When we minimize free energy, we can either lower the total energy, increase the amount of unusable energy, or both. In self-assembly, all of these possibilities come into play.

One of the most familiar and most stunning examples of self-assembly is the aforementioned snowflakes. Snowflakes are crystals of water ice. If you ask my students why snow crystals form, they would answer that it minimizes free energy! And, indeed, as we saw in Chapter 3, there is a critical temperature below which the lowering of the free energy leads to the formation of snow flakes, while above that temperature, free energy is minimal when water remains liquid. Yet, we would like to have more information. Snowflakes are such beautiful, intricate structures, and explaining their structure as a reduction in free energy seems like a cop-out. The result is a reduction in free energy, but how does it happen? How does the structure form?

The creation of snowflakes is a perfect example of the combination of chance and necessity, entropy and energy. The six-sided symmetry of snowflakes is a macroscopic representation of the underlying molecular symmetry of water ice. Water molecules like to arrange themselves in a hexagonal (six-sided) pattern, like a three-dimensional puzzle. Different external conditions of moisture, temperature, and pressure influence the growth of the crystal, leading to different branching patterns. Since all six sides of the growing crystal are subjected to the same conditions at any given time, they grow more or less (although not exactly) in the same pattern, creating the beautiful symmetry of a snow flake.

The growth of a snowflake shows how complexity and beauty can arise from cold physical principles (no pun intended). Freezing is accompanied by a release of heat. This heat has to be removed (increasing the entropy of the environment) to allow the water to freeze. Thus ice crystals will grow fastest at those places where heat is removed the fastest, namely, at the endpoints of any spiky features. Spikes, however, are already the result of rapid growth. This creates positive feedback: The locations on a crystal that grow the fastest become spiky, which allows better heat transfer from these locations, which makes them grow faster, and so on. The result is what physicists call an instability, leading to the formation of long, spiky features. But there is an opposing force as well: The local temperature also depends on the spikiness of the crystal; spikier parts (those with higher curvature) melt at lower temperatures, thus reducing the temperature difference between spiky parts and the environment. This reduction in temperature difference slows heat transfer, counteracting the effect of the exposure of the spikes. Once a spike becomes too pointy, growth will slow down—a negative feedback that counteracts the positive feedback. When this happens, the crystal develops branches. The combination of these two tendencies, to become spiky and to develop branches, leads to the dendritic growth of snowflakes.

Another important aspect of snow crystal growth is *mass transport*—how new water molecules can reach the growing crystal and become part of it. This is mostly a question of diffusion, the random motion of water molecules in the vapor of clouds. A crystal can only grow as fast as water molecules can reach it. The crystal's growth is *diffusion limited*. Physicists call such growth diffusion-limited aggregation. Again, parts of the crystal that stick out are easiest to reach by random diffusion (think of lightning rods, which are easiest to reach by the random motion of a lightning bolt), which adds to instability.

How complicated a simple snow flake is! The example of the snowflake illustrates how the interplay of energy, entropy, and various mechanisms (diffusion, heat transfer), together with an underlying geometry, can create complexity and variety, but also a certain robustness of structure. Snowflakes are not alive, but they serve to illustrate that complex structure formation, as encountered in living things, can arise from dynamic physical laws. Moreover, snowflakes are nonliving examples of structures that exhibit both repetition (all snowflakes are six-sided, spiky, and branchy) and nearly unlimited variety within this basic pattern.

Nanotechnologists use similar principles to create ever-more-complicated nanostructures. The ultimate goal is to design circuits or nanoscale machines that grow themselves, molecule by molecule. The simplest case of self-assembly occurs when spherical particles are deposited on a surface. If they have some attractive forces between them and some way to move around until they find a nice place to settle down (typically where they have one or more neighbors), they inevitably form close-packed layers. A close packed layer looks like stacked oranges: Place (equal-size) oranges next to each other, so they all touch, and you end up with a close-packed, hexagonal layer. This is the most common structure found in nature (a honeycomb is a good example), because it maximizes contact between neighbors and wastes a minimum amount of space.

But hexagonal close-packing gets boring very quickly. If we want to make more complicated structures, we have to start with more complicated building blocks and with more innovative ways of putting these building blocks together. There are various ways to cajole molecular building blocks into forming complex structures. Use oddly shaped molecules that only attach to other molecules at certain attachment points (chemists call this directional bonding), or use nonequilibrium conditions that somehow bias the structure to form in a certain way (e.g., apply electrical fields, pressure, or a liquid flow), or let entropy give you a helping hand.

Let's Cooperate

Mayonnaise is not good for you, but it sure tastes good on a sandwich. While it contains water and oil, it is remarkably stable and uniform, even though we know water and oil do not mix. If you've ever made mayonnaise (I tried once, but it wasn't edible), you know the secret is that mayonnaise is made with egg yolks. But why would eggs stabilize an otherwise unstable mixture of oil and water?

Egg yolks are mostly water, with some proteins and a sizable amount of fat. The fat part consists of fatty acids (oils), lipids, and other compounds. Lipid molecules have two important parts: a salt *head* (which is water-soluble, or hydrophilic) and a fatty acid *tail* (which is water-insoluble, or hydrophobic). The molecules are a kind of conjoined twin, with each part having different affinities. In a mixture of oil and water, lipids can satisfy each part of their split personality. The hydrophilic head can stick out into water, while the hydrophobic tail can immerse itself in the oil. Lipids are just one example of such *amphiphilic* molecules (i.e., they are soluble in water *and* oil). Other examples are detergents, soaps, and other surfactants. Because of their dual solubility, these materials make oily, greasy stuff water-soluble and enable us to wash our dirty dishes.

This sounds good, but there is a problem: Oil and water molecules are always pushed around by the molecular storm. The unfortunate lipid or detergent molecule cannot enjoy its divided loyalties for long, before the oil and water mix again. How can these molecules get around this problem? They find strength in numbers. When they come together in sufficient numbers, they can form stable structures—*micelles*—spherical protective shields that completely engulf droplets of oil and permanently separate the oil from the surrounding water. Micelles are made of many, many lipid molecules. These structures allow the lipids to satisfy their conflicted hydrophilic-hydrophobic affinities while keeping the adversaries, oil and water, apart. The reduction of free energy associated with this arrangement is so large that the thermal motion of the molecular storm is too weak to break the micelles. The emulsion of oil and water has become stable. And that is why mayonnaise does not separate.

Micelles can also form in the absence of oil. If we place a single amphiphilic molecule into water, it would have no choice but to expose its hydrophobic tail to the water. However, if there are enough molecules, they can form micelles, namely, spheres with an outer layer of hydrophilic heads and an inner core of hydrophobic tails. With the hydrophobic tails safely tucked away, this arrangement reduces free energy considerably. The formation of such micelles is a textbook example of cooperative dynamics. Cooperative dynamics, or cooperativity, plays a central role in structure formation in living systems. In a nutshell, cooperativity means that a structure cannot form unless a certain number of molecules "cooperate" to form the structure. In micelles, this means that a micelle requires a minimum number of lipid molecules to form. You need a certain number of molecules to form a complete sphere with a rather close-packed core of hydrophilic tails. Too few molecules, and you cannot form a closed sphere. Too many, and the sphere becomes too large, leaving a void in the center, which costs energy. Thus there is an optimal number of lipids in each micelle.

Imagine how a spherical micelle could form gradually. First two amphiphilic molecules meet and point their hydrophobic tails at each other. Then, they are joined by a third and so on, until the molecules form a closed sphere. But this is not at all what happens. Measurements show that below a minimum concentration of molecules in solution, no micelles form. The molecules remain lonely hearts. However, as more lipids are added, suddenly complete micelles start to form. Why is that? Micelles do not form gradually. Rather, at the critical micelle concentration, the probability that the correct number of molecules will spontaneously meet becomes large enough to allow the immediate formation of complete, stable micelles, without any partially assembled micelles along the way. Micelle formation is an all-or-nothing proposition. Either enough lipid molecules meet at the same place and cooperatively form a micelle, or no micelles form at all. It is like a proper soccer match: Either you (initially) have twenty-two players on the soccer field, or you don't have a game.

One result of this all-or-nothing dynamic is that there is a very sharp change in the properties of a lipid solution when it reaches critical concentration. One such property is the *osmotic pressure*, that is, the pressure exerted on a wall due to the random motion of dissolved molecules (remember, pressure is just the sum total of all impacts of randomly moving molecules with a wall). In his famous papers on Brownian motion, Einstein showed that the osmotic pressure does not depend on the size of the molecules or particles in solution, but depends only on their concentration and temperature. Thus, when micelles suddenly form from a large number of single molecules, the total number of particles in solution is greatly reduced. Where we originally had many lonely lipids, we now suddenly have just a few micelles—each counting as just one particle. This reduction in the number of free particles in solution leads to a sharp reduction in osmotic pressure. Such sharp, all-or-nothing transitions are a signature of molecular cooperativity.

Cooperativity, the observation that some structures form only when a minimum number of molecules cooperate, is ubiquitous throughout biology. Molecules do not cooperate in the sense that people cooperate. They are, after all, molecules. Molecules are pushed around randomly by the molecular storm. For example, if lipids can form a micelle, they will reduce their free energy enormously. If the concentration is too low, you can never get enough molecules together at the same location. But at some critical concentration, when you have enough molecules around, they find each other in sufficient numbers, and micelles form spontaneously.

Lipids can also form more complicated cooperative structures in water: double-walled spheres, called vesicles. Inside the double wall, the hydrophobic tails are safely tucked away from the surrounding water, while the two surfaces, one on the outside and one on the inside of the sphere, face water molecules (Figure 4.1). A vesicle separates one volume of water from another. If we place different chemicals inside the vesicle, we create an isolated, nanosize reaction chamber. All kinds of interesting things could go on inside this tiny chamber—life, for example. Indeed, the walls surrounding living cells are made of lipids; a living cell is in some sense a giant self-assembled vesicle with a lot of really complicated chemistry going on inside.

Entropic Force

At the nanoscale, the molecular storm reigns supreme. Yet, the random thermal forces of the storm paradoxically lead to the creation of ordered structures. As a matter of fact, without it, nothing would be happening at all! Self-assembly requires that the pieces that make the structure are shuffled around until they find a comfortable resting place where they mini-

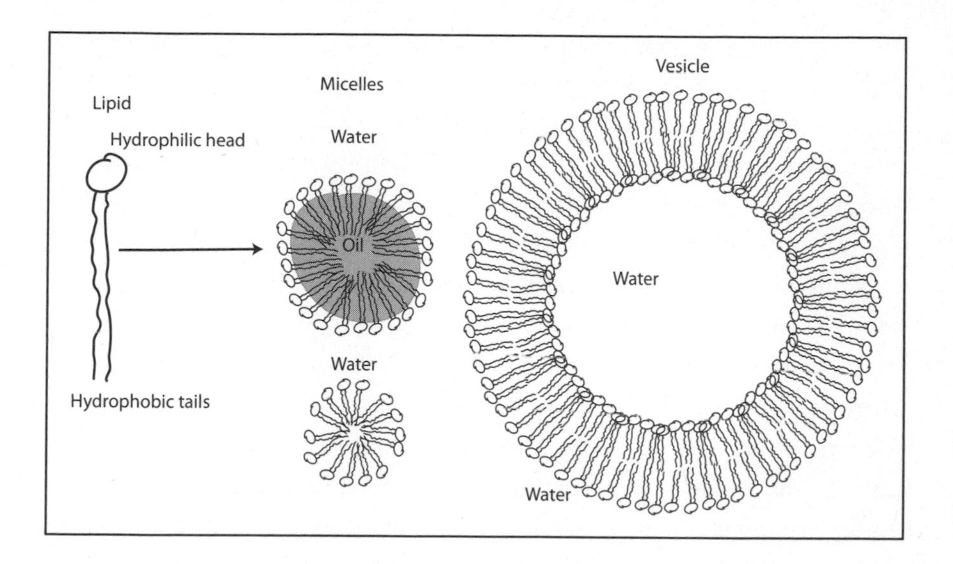

FIGURE 4.1. Lipids are amphiphilic molecules with a hydrophilic head and two hydrophobic tails. In a water and oil mixture, lipids can surround small oil droplets and form micelles. Lipids can also form micelles in the absence of oil, but only if a specific number of lipids comes together to form a spherical shell. This is an example of cooperativity. Usually, however, lipids prefer to form vesicles, which are structures made of a double wall of lipids, separating one volume of water from another. Vesicles form the basis of cell membranes and are used inside cells to transport chemicals around.

mize their free energy. As we have seen in Chapter 3, structures form sometimes to maximize random motion, that is, to maximize entropy.

Cells are incredibly crowded places, stuffed full of large proteins, DNA and RNA molecules, sugars, lipids, ions, and innumerable water molecules. It has been estimated that the average space between proteins in living cells is less than ten nanometers. Since proteins are between ten and one hundred nanometers in size, this is equivalent to a crowded parking lot with just a foot or less between each car. When things are this tight, it becomes tricky to maneuver past each other. In addition to this crowding, every space between proteins is filled with water, ions, sugars, and other assorted small molecules. You now have an idea of what a crowded mess a cell is. This crowding has consequences, many of which are not well understood, because when we do experiments on proteins, the steps are usually conducted outside the cell in a test tube, where proteins have plenty of space.

One surprising consequence of crowding is that the small guys (water, ions) rule over the big guys (proteins, DNA, lipid vesicles). Imagine two large protein molecules approaching each other in a sea of smaller molecules. The small molecules, because of their finite size, can only approach the big molecules so far. If they get any closer, they'd have to press into the large molecule. Thus, every large molecule is surrounded by an exclusion zone (also called a depletion zone) that cannot be penetrated by small molecules and thus leads to a reduction in total space available to the small molecules. It is a kind of molecular personal space: Imagine our large molecules as adults standing around a room, each with their own personal space, and the small molecules as kids running around. It is difficult for the children to run around, as they have to avoid entering the adults' personal spaces. Suddenly, the adults pair off, and couples start to embrace. What previously were two personal spaces, one for each person, becomes one merged personal space per couple. Suddenly, more room is available for the kids to run around, and they promptly take advantage of their newfound freedom (Figure 4.2).

The same can happen in our molecular world. If two large molecules stick to each other, their exclusion zones merge, and more space is available for the small guys. More space means more available microstates and thus increased entropy. Thus, the increase in *order* due to the binding of the large molecules is more than paid for by the increase in *disorder* arising from giving the surrounding smaller molecules more space to move, increasing the small molecules' entropy. What we end up with is a force that is not the result of decreasing energy, but of increasing entropy. Such strange forces, which can only exist at the molecular scale, are called entropic forces.

Cells take advantage of these entropic forces to assemble a variety of molecular structures, including collagen, actin, and microtubule filaments. Collagen is the main ingredient of the extracellular matrix, the web of material outside our cells that holds our bodies together. Actin and microtubules are fibers that form the malleable skeleton of our cells.

How do we know biological fibers such as collagen are assembled by entropic forces? Remember the stock answer to any question of why something happens: because it minimizes free energy. Free energy, we have learned, is energy minus unusable energy. Unusable energy is given by the product of temperature and entropy. Since temperature comes into

FIGURE 4.2. If large molecules are separate from each other, their exclusion zones make it difficult for small molecules to move around. Once the larger molecules bind to each other, however, their exclusion zones overlap, creating more space for small molecules. This increases the entropy of small molecules and leads to an entropic force's driving large molecules to bind to each other.

the picture, we can do a relatively simple test to see if an assembly is driven by a reduction of energy or by an increase in entropy. If the assembly proceeds faster when we increase the temperature, it is mostly driven by a reduction in energy. On the other hand, if the assembly is *slower* when we increase the temperature, entropy is the culprit.

Most chemical reactions proceed faster at higher temperatures. Try to dissolve sugar or salt in ice water, and then try dissolving these substances in hot water, and you'll see what I mean. Sugar dissolves much faster in hot water than in cold, indicating that the dissolution of sugar is driven by a reduction in energy. In the self-assembly of collagen, however, the reaction *slows down* when the temperature is increased. During collagen formation, total energy (E) actually *increases* when single collagen units combine to form fibers. Processes in which the energy of the system is increased should be unfavorable. But at the same time, the entropy (S) also increases. This increase in entropy more than trumps the unfavorable energy increase, leading to an overall reduction in *free* energy ($F = E - TS$), because the entropy term (TS) is subtracted from the energy.

There are numerous biological examples of such entropy-driven self-assembly processes. In addition to collagen, actin, and microtubules, the self-assembly of viruses and the flagella of bacteria are entropy driven. These examples show that order can be created while increasing entropy. Entropy, as we have mentioned, is often confusingly equated with disorder. Well, it *is* a kind of disorder, but what we overlook in such a simplified picture is that the creation of disorder in one part of the system can be coupled to the creation of order in another part of the system. Just ask a snowflake.

I Hate You; I Love You

Soap, detergents, and lipids are all examples of linear molecules that have one hydrophilic and one hydrophobic end. But what makes molecules hydrophobic or hydrophilic in the first place?

We know that oil and water don't mix (unless you add eggs). Why? Water is a peculiar liquid, quite unlike any other liquid in the universe. For example, compared with liquids of similar molecular construction, water has to be heated to a much higher temperature to melt or boil. It also has the curious property that its solid form (ice) is lighter than its liquid form; this is why ice cubes float on your drink, rather than sink. For most other substances, the opposite is true: The liquid is lighter than the solid. All these strange properties of our favorite liquid are mainly due to hydrogen bonds. Hydrogen atoms of water molecules form strong bonds with oxygen atoms of neighboring water molecules, forming "bridges" between the molecules. Because these hydrogen bonds in liquid water are stronger than we would expect, water does not evaporate as easily as other liquids.

Hydrogen bonds arise because in each water molecule, which is made up of one oxygen atom and two hydrogen atoms, the oxygen "steals" the hydrogen's electrons. The leftover hydrogen ions are positively charged, while the oxygen atom, now having two excess electrons, is negatively charged. When two neighboring water molecules come close to each other, the positively charged hydrogen on one molecule is attracted to the negatively charged oxygen on a neighboring water molecule. In liquid water, these hydrogen bonds form and break continuously at a high rate. They are but fleeting couplings. Yet, they make a profound difference for the stability of the liquid and lead to its high boiling temperature. In ice, the hydrogen bonds are essentially permanent. They are also *directional*, that is, water molecules can only form hydrogen bonds in ice if the molecules are arranged in a certain way. This arrangement leaves large, molecule-sized voids in the ice structure, making ice less dense than liquid water. Hence, ice floats on water.

Nonwater molecules can also form hydrogen bonds with water, as long as they have some negatively charged parts. This includes many acids, sugars, and alcohols (a good thing—it would be a shame if alcohol and water didn't mix). Other molecules, while not forming hydrogen bonds, are still quite happy in water, because they can accommodate their charges by surrounding themselves with water molecules. This is what ions do, and it explains why salt dissolves in water. Oils do not have any charges; they are a neutral bunch of molecules. Thus water cannot form hydrogen bonds with oil; nor are there any charges to accommodate. Oil is a molecular party-pooper—not very sociable. If oil is placed into water, water molecules form a kind of cage around an intruding oil molecule, trying to maximize hydrogen bonding to their own kind. The cage structure they form induces order in the water and therefore decreases entropy. A decrease in entropy leads to an increase in free energy, which is a bad thing. Trying to put oil into water comes at high (entropic) cost. What's an oil molecule to do? Get out of the water and find its own kind. Soon enough, oil molecules form droplets, and eventually the droplets coalesce and leave the water all together. Oil and water separate.

The hydrophobic force, which is responsible for the separation of oil and water, is therefore another example of an entropic force. While the hydrophobic force separates oil from water, it also provides a powerful drive for oil molecules to come together and coalesce. In water, hydrophobic molecules attract each other. This attraction of hydrophobic molecules is an important mechanism for self-assembly. One tasty example of this is cheese. Cheese can be made by adding salt or acid to milk or, in a more modern approach, by adding rennin. Each method leads to essentially the same result, but interestingly, the physics of how each works is different. About 80 percent of the protein in cow's milk is in the form of casein, a phosphorus-containing protein. Casein molecules are hydrophobic, but also negatively charged. Their hydrophobic nature means that if they could get close enough to each other, the hydrophobic force would make them stick to each other. However, their negative charge keeps them apart. Now comes the cheese maker. Adding salt provides free positively and negatively charged ions to the milk. The positively charged ions crowd around each negatively charged casein molecule and shield its charge. Now casein molecules can approach each other and start to stick, forming cheese.

How about acid? Acid increases the amount of hydrogen ions in the milk, changing the milk's pH. The positively charged hydrogen ions stick to the casein and neutralize it. Again the casein molecules start to stick and cheese is made.

Adding salt or acid, of course, affects the taste of the cheese, making it taste salty or sour. Sometimes this is not desirable. This is where rennin comes in. Rennin is an enzyme, a protein that promotes a certain reaction that cuts off the charged part of the casein. Without its charged part, the casein molecules are quickly forced together by the hydrophobic force.

The importance of hydrophobic forces is not limited to the kitchen. Hydrophobic forces play a crucial role in shaping the very molecules that keep us alive, as we will discuss in the following section.

Molecular Origami

A life-or-death example of a molecular self-assembly process is the folding of large biological molecules, especially proteins, into their active shapes. Proteins are made of long chains of chemical units called amino acids, strung together like pearls in a necklace (Figure 4.3). On Earth, every protein in every living organism uses just twenty different amino acids (there are many more amino acids that could in principle be used, but nature has decided to use only these twenty). The particular sequence of amino acids in each protein is encoded in the cell's DNA. Different amino acids have different affinities, courtesy of the hydrophobic force. These tendencies to attract or repel determine the particular, coil-like shape of a protein. Get the folding of a protein wrong, and you could end up with Alzheimer's disease, mad cow disease, or cystic fibrosis—examples of diseases in which

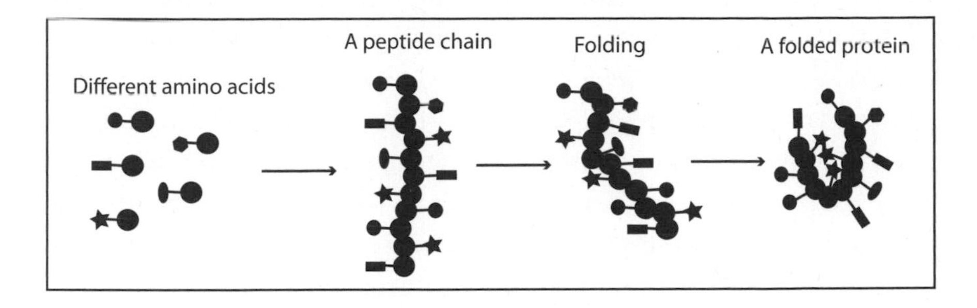

FIGURE 4.3. Amino acids are molecules with identical backbones, but different side chains. The backbones can link together by peptide bonds, forming a peptide chain. This chain is pushed around by the molecular storm until it folds into a stable structure with hydrophobic side chains in the center (stars) and hydrophilic ones on the outside.

certain proteins fold into nonfunctional or even dangerous shapes, as a result of mutations or other unknown reasons.

How does a protein find its optimal shape? When a protein is made by the molecular machinery of the cell, it initially comes out as a long string of amino acid beads. Some of these beads are destined to bind to each other, but initially the amino acid string wiggles around aimlessly, pushed by the molecular storm of bombarding water molecules (Figure 4.3). A molecule does not "know" how to fold—it needs to find its shape through random changes in its shape.

Even a small protein can fold into innumerable possible shapes, the number of which far exceeds the number of atoms in our universe. The energy of the protein depends on all the twists and turns of its chain and the contacts between amino acids along the chain. If we were to measure the twists of the chain (by assigning angles to each link of the chain), we could calculate the energy of the protein for any possible shape. Some shapes would have larger energies, some lower. The greatest number of amino acids bind together in the lowest-energy shapes. As shown in Figure 4.3, in the lowest-energy configuration, the hydrophobic amino acids are safely tucked inside, while the hydrophilic amino acids are found on the outside of the coil.

Imagine a map representing the energy of any possible shape of a protein. The coordinates of the map (north, west, south, and east) represent the angles of the links, and the topographic height, at each coordinate, represents the corresponding energy. This creates what physicists call the energy landscape of the protein. Figure 4.4 shows possible shapes of a short protein chain with the associated energy shown as the height on the landscape. The protein shown in this schematic has just four links, while real proteins consist of thousands of amino acid links, making the energy landscape fantastically complicated. The different links of the protein are randomly pushed around by the molecular storm, causing the protein to continuously change shape. This ongoing change in shape is represented by the white path across the landscape. Each location along the path corresponds to a specific shape of the protein (some examples are shown in the figure).

Sometimes the molecular storm pushes the protein into an uncomfortable, high-energy shape (a hill in the landscape); quickly, the protein relaxes down the hill to a more comfortable configuration. When the protein shape approaches the optimal shape, its energy reduces and it falls into an energy funnel in the landscape. Further reduction in energy draws the protein to the bottom of the funnel, where it assumes its lowest-energy configuration and optimal final shape.

As the molecular storm pushes the protein around, the succession of different protein shapes can be represented as a meandering path across the (abstract) energy landscape (the white path in Figure 4.4). Physicists talk about a diffusion across the energy landscape (diffusion here means random motion). Without the molecular storm, the molecule would be stuck in a fixed shape and would never find the optimal shape that minimizes its free energy. The diffusion across the energy landscape is a search that eventually leads the protein molecule to its optimal, lowest-energy shape. This optimal shape of the protein determines the function of the protein. Without its three-dimensional shape, a protein is nothing. The particular sequence of amino acids making up the protein has meaning only as far as it determines the protein's 3-D shape.

For some proteins, the search for the optimal shape would simply take too long. For such proteins, life invented smaller proteins that chaperone, or guide, the larger molecule toward its proper shape. These *chaperonin molecules* serve as local molds that preshape parts of a protein and help it find its way along the energy landscape more quickly.

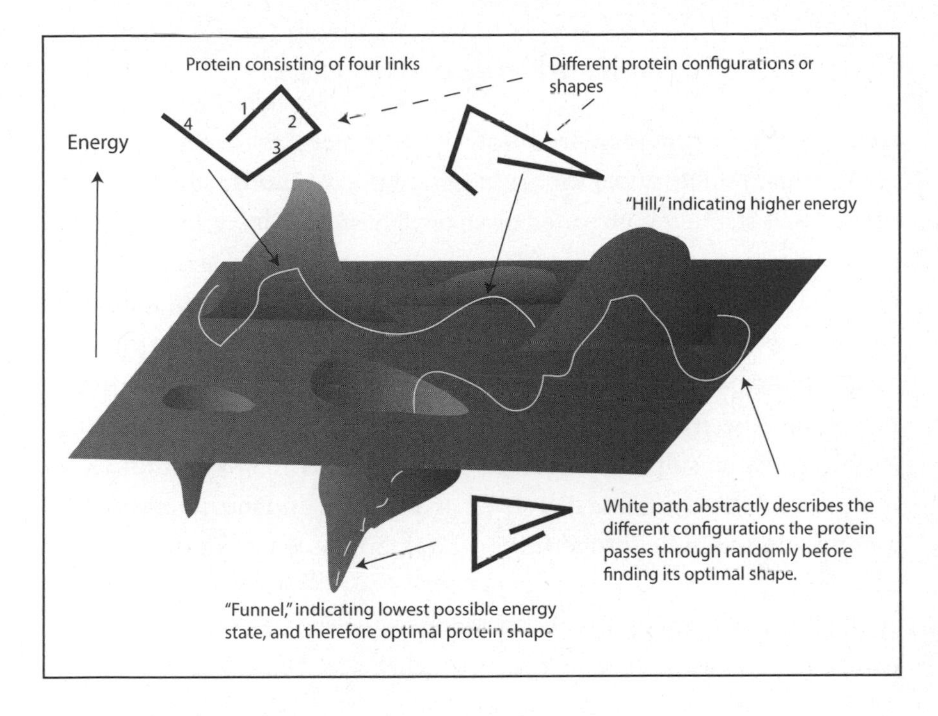

FIGURE 4.4. The energy landscape of folding a protein.

Protein folding is possibly the best example of how physical laws, ran domness, and information—provided by evolution—work together to create life's complexities. The amino acid sequence of a protein is determined by the cell's DNA, according to the genetic code. This information evolved over billions of years. But the information in our DNA *only* encodes the amino acid sequence; it does not encode the final 3-D shape of the protein. The 3-D shape is a result of the energy landscape, which is determined by physical forces (hydrophobic forces, electrostatic forces, bending energies, etc.) acting on the particular sequence of amino acids. This shape also depends on external conditions (pH, temperature, ion concentration). Thus a large part of the necessary information to form a protein is not contained in DNA, but rather in the physical laws governing charges, thermodynamics, and mechanics. And finally, *randomness* is needed to allow the amino acid chain to search the space of possible shapes and to find its optimal shape.

Schrödinger's Riddle

Erwin Schrödinger wondered how life's molecules, especially DNA, could remain stable under the onslaught of the molecular storm. He postulated that stability is provided by strong chemical bonds. He was partly right—there are strong chemical bonds along the backbone of both DNA and proteins—but he missed the most interesting point: The functional structure of the large molecules in cells is not held together by a few strong bonds, but by many relatively weak ones, especially hydrogen bonds, but also hydrophobic interactions and so-called disulfide bridges (bonds between sulfur atoms attached to sulfur-containing amino acids in a protein). If the unique coil shape of a protein and the double helix of a DNA strand are held together by weak bonds, why are these shapes so stable?

If all bonds in the molecules of living organisms are too strong, there can be no motion or change. If they're too weak, there can be no stability. The key is to find a middle ground between stability and flexibility. If things are too stable, they cannot function—imagine a car with lead wheels. If they are too flexible and fragile, they cannot function either. The solution that life has found is to capitalize on the cooperativity of many bonds. As we saw earlier, many weak bonds cooperating make a molecule very strong, while individually, the bonds can easily be broken, allowing for rearrangement when needed.

How can bonds be cooperative? In the aforementioned micelle, molecules act cooperatively to form a larger structure. If we replace *molecule* with *bond*, we have a similar situation. Once a DNA molecule is formed, the two strands of the DNA are held together by numerous hydrogen bonds. A single hydrogen bond can easily break at room temperature. The average lifetime of a hydrogen bond in water at room temperature is in the picosecond range (that is, a thousandth of a billionth of a second). However, if we have many hydrogen bonds working together, the spontaneous separation of two DNA strands would require a simultaneous, *cooperative* breaking of a large number of bonds—and this will not happen spontaneously. Single bonds may temporarily break in a DNA molecule, but they will quickly reform—forced back together by the presence of neighboring hydrogen bonds, which keep the DNA molecule in shape. In living cells, DNA retains its double-helix structure when the molecule is not in use. However, when the information contained in DNA needs to be read out, the double helix must be opened up. This is achieved by specialized proteins, which locally lower the barrier to undo a number of hydrogen bonds and, like a zipper, unzip the DNA where needed and then close it back up (see Chapter 7 for more on this).

Stability in DNA is also actively achieved by repair enzymes. If for some reason a part of the DNA molecule becomes corrupted, the repair enzymes will fix the mistake, using an adjacent section of the molecule that is undamaged as a chemical template to repair the corrupted part. The stability of DNA is therefore the result of bond cooperativity, complementarity of the two strands, and active repair by the cell's machinery.

The situation is similar for proteins. A protein's shape is held together by hydrophobic, hydrogen, and disulfide bonds, again cooperatively stabilizing the shape of the protein. The protein's particular amino acid sequence is chosen by evolution such that a protein is just flexible enough to do its job (we will find out in the next chapters what kind of jobs they do) while keeping its overall shape. If a protein does get out of shape, the cell quickly discards it and recycles it back into its components.

The Emergence of Long Time Scales: Forcing Water Molecules to Cooperate

Cooperativity is almost inevitable when you have many entities interacting with each other. I found this out in my own research when my students and I looked at simple liquids—water or oil—squeezed down to a few nanometers. While solids, in the form of crystals, have long-range order (atoms are arranged in an orderly pattern over large distances throughout the crystal), liquids have short-range order: If you were sitting on a water molecule in a pool of liquid water, you would see neighboring molecules at an average distance of 0.25 nanometer (nm). These neighbors would move around, but on average, there would be a cloud of water molecules surrounding you at this distance. Beyond this distance, the next-nearest neighbors would feel the presence of your neighbors, and you would see an excess of molecules at an average distance of 2×0.25 nm =

0.5 nm. However, because of the incessant motion of all molecules, this excess of molecules at 0.5 nm would be a little bit more smeared-out than the excess of molecules at 0.25 nm. Going even further, there would be an even more smeared-out excess at 0.75 nm and so on, until at a distance of five to six molecular diameters, thermal motion would have smeared out any semblance of order, and water molecules could be found with equal probability at just about any distance.

In my laboratory, we are using a different method to probe the short-range order of liquids. When liquids encounter a solid surface, they arrange themselves into molecular layers. The first layer, closest to the surface, is well defined. There is a high probability that we find water molecules at 0.25 nm from the surface. The second layer is a little bit more diffuse, and as we move away from the surface, the layers become more and more smeared-out, until order vanishes altogether after about five or six layers, as discussed above. This layering can be measured by an AFM. In our lab, we immerse an AFM cantilever and probe tip into water and then slowly move the tip toward the surface. To measure the response of the intervening water, we oscillate the tip at an amplitude (size of oscillating motion) of less than 0.1 nm—the diameter of a single hydrogen atom! Monitoring the amplitude, which changes slightly as we push through the water layers, we measure the "springiness" of the water layers (or "stiffness," as we call it), that is, the capacity of water to push back at the tip if we push the tip against it. The stiffness fluctuates up and down with a period of 0.25 nm—the size of the water molecules—illustrating the molecular graininess of the liquid. An example of such a measurement is shown in Figure 4.5.

In 2006, my post-doc Shiva Patil and my student George Matei were performing such measurements on a silicone oil, octamethyl-cyclo-tetrasiloxane. OMCTS has large molecules (if you are a nanoscientist), with diameters of 1 nm. Together with the stiffness, we also measured the damping of the liquid layers. While the stiffness tells us how much the liquid layers spring back into their original shape (i.e., how much energy they store), the damping tells us how much energy is lost as we push on the layers. The damping also fluctuated with the spacing between the tip and the surface, with the same period as the stiffness. Looking at the data, Shiva noticed something peculiar: The damping fluctuation wasn't always

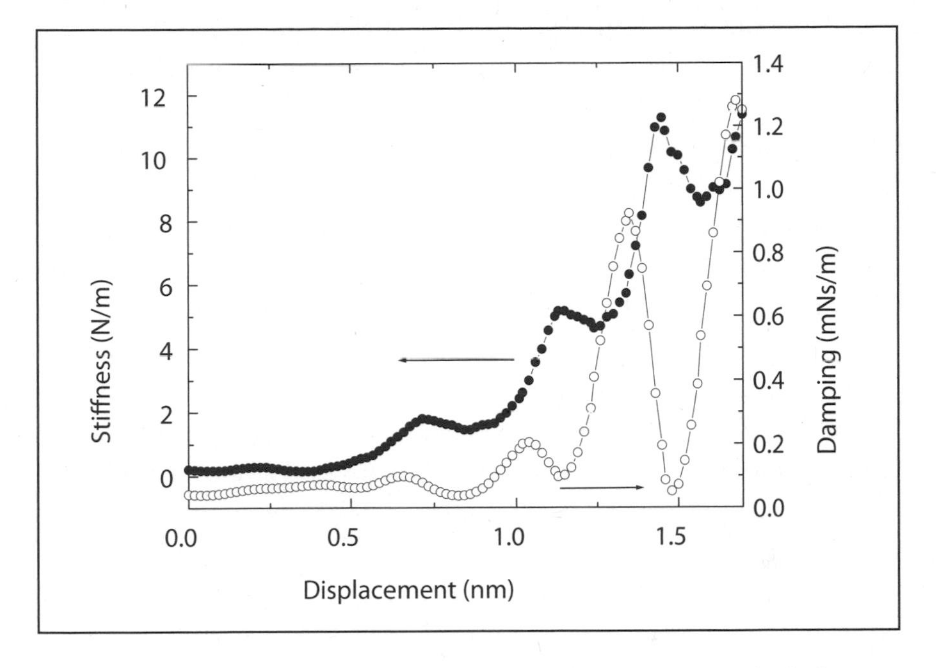

FIGURE 4.5. Measurement of the stiffness and damping of molecule-thick water layers, using an atomic force microscope (AFM). Each peak in the stiffness (filled circles) corresponds to a thickness of water between the AFM tip and a flat surface, which contains an integer (1, 2, 3, . . .) number of water molecules. The fact that the stiffness peaks and the damping peaks (open circles) don't line up means that the few-molecules-thick water layer responded like a solid to being squeezed by the AFM tip.

in line with the fluctuations of the stiffness. Sometimes the peaks in the damping lined up with the peaks in the stiffness, but sometimes they were switched—peaks in damping coincided with valleys in the stiffness. What was the difference between the measurements? We looked at everything—ion concentration, liquid purity, problems with the instrument. After two weeks, Shiva found that the difference between the measurements was in the speed at which the tip approached the surface. Yet, we were moving the tip so slowly in our measurements—the fastest speed we used was 1.2 nm per second—that it shouldn't have mattered. At this extremely slow speed, it would take almost a second to push the tip through one layer of molecules. For liquid molecules at room temperature, a second might as well be an eternity. Thermal motion is much, much faster than this. Yet, we found that something quite dramatic happened when we went from a speed of 0.4 nm per second to 0.6 nm per second and beyond. At the slower speeds, damping and stiffness fluctuated together. This meant that the liquid behaved like a liquid—it just got thicker and thinner as we pushed the tip through. At faster speeds, the liquid assumed a new state; it became springy with very little energy loss. Usually, only solids could do this. Since then, we have found the same effect in liquid water, although here we have had to squeeze a little bit faster to make it happen. Water is still liquid at 0.8 nm per second and starts behaving like a solid beyond that speed. The transition between the two states of the liquid is extremely sharp: Below a critical speed, it always behaves like a liquid, and within a very small range of speeds, it completely switches to a solid-like response.

The same thing can happen for bulk liquids (i.e., liquids that are not confined to nanometer spaces), but only at very fast speeds. Our finding was that at the nanoscale, these dramatic changes happened at very, very slow speeds. But how is that possible? Molecules in a liquid move at incredible speeds, much faster than our AFM tip. They should have plenty of time to accommodate the approaching tip, and the liquid should remain liquid. The answer is—you guessed it—cooperativity. Once water is squeezed to just a few molecular layers between the surface and the tip, it becomes difficult for the water molecules to move out of the way of the approaching tip. Due to the confinement, the molecules don't have the freedom of motion they enjoyed in the bulk. When confined to the nanoscale, many molecules have to move in concert to create a hole into which the tip will move—water molecules have to cooperate to move out of the way. What is truly remarkable about these results is that it takes water molecules extremely long, of the order of seconds, until they *randomly* happen to move in a coordinated manner to create a hole. That is a million billion times longer than the average time between water molecule collisions. A crude calculation indicates that thirty to forty molecules—not very many—would have to be involved in the collaborative motion to create such a long time scale. Cooperativity can create not only sharp transitions, but also large changes in time scales, making even molecular processes take seconds or even minutes.

Molecular Switching

Amphiphilic molecules, such as lipids and detergents, suddenly form micelles once they reach a critical concentration. This results in a sharp change in osmotic pressure. In our confined-water experiments, we saw a rapid switching from liquid-like to solid-like behavior in response to a small increase in squeeze rate. The occurrence of rapid changes at critical values for certain control parameters (concentration for micelles, speed for confined water) is a signature of cooperative behavior.

Cooperative behavior is not restricted to molecular biology. It is a ubiquitous, but underappreciated facet of our world. For example, some economists have argued that the financial crisis of 2008 and others before it, including the Great Depression of the 1930s, were the result of a cooperative failure of banking. This is how it goes: There is always a low rate of bank failures. As long as that rate is low, the overall financial system is relatively unaffected. But if banks fail at a rate exceeding a certain critical threshold, the interconnectedness (cooperativity) of banks pulls the whole market into the abyss. Thus, what looks like another, albeit somewhat larger, fluctuation in the financial market suddenly leads to rapid, profound change. This is cooperativity at work.

Cooperativity leads to *switching*. Yesterday everything was fine; today the economy has crashed. A second ago, lipid molecules were happily floating around, and then, suddenly, they form micelles, and osmotic pressure drops precipitously. Water acts like a liquid, but squeezed above a slow critical rate, it suddenly bounces like rubber. A DNA double helix was fine a second ago, but suddenly it catastrophically unzips when it is heated. The San Andreas fault was quiet an hour ago, and now all hell breaks loose. We should not underestimate cooperativity!

Despite some of the negative examples of its effects, cooperativity is crucial for the function of living cells. Changes in a molecule's shape are driven by cooperativity of many bonds, and the shape change, often sudden and dramatic, can be driven by a relatively small external change. This behavior allows for the creation of *molecular switches*, molecules that can effect large changes in response to small causes, such as the binding of a small molecule. This in turn allows the creation of *molecular circuits*, which control the activity in a cell. In electronics, a transistor is an element that allows a small change in a voltage to control a large current. Transistors are electronic switches, and they are the root of modern electronics, from radios to computers. Similarly, molecular switches in cells serve as control units to make cells work. They work on the principle of cooperativity, which in turn is made possible by the use of many small bonds, many of them of entropic origin.

All Energies Are Created Equal—At Least at the Nanoscale

Thermal motion, entropic forces, and cooperativity—some of the strange properties of nanoscale systems—are important for our understanding of life at the molecular scale. Another, truly astonishing property of the nanoscale is *the* key to understanding how the coordinated activity of cells is generated. This property relates to how energy is transformed from one form to another. One of the most astonishing features of life is that living beings can take energy from food and turn it into directed motion. Past generations attributed this magical feat to life forces. However, the continued search for physical explanations has brought scientists to the molecular scale. Proteins, DNA, RNA, and other large molecules inside cells seem to be the fundamental functional units that make the cells work. Some of these molecules must be able to convert energy from one form to another; they must be acting like *molecular machines*.

Machines are energy-conversion devices—a car engine, for example, converts chemical energy into kinetic energy. However, a car engine sitting in a pool of gasoline with no connections would not jump to life. Yet, the molecular machines in our cells do just that: Pluck a molecular machine, such as myosin, out of a cell, give it some chemical fuel (called ATP in cells), and it will "come to life." Molecular machines are autonomous machines. Why can molecular machines work autonomously, while our familiar, macroscopic machines cannot?

It turns out there is something very special about the nanoscale when it comes to converting different forms of energy into each other. Intriguingly, *only* at the nanoscale are many types of energy, from elastic to mechanical to electrostatic to chemical to thermal, roughly of the same

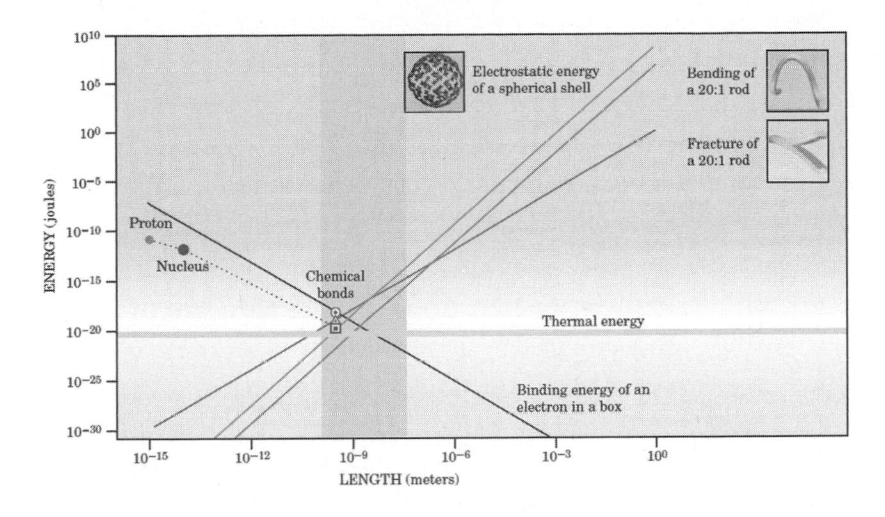

FIGURE 4.6. Electrostatic energy, chemical bond energies, and elastic energies all converge at the nanoscale (10^{-9} meters), where they meet thermal energy (the molecular storm) at room (or body) temperature. This confluence of energy scales at the nanoscale explains self-assembly and the possibility of molecular energy conversion devices and machines. Reprinted with permission from Rob Philips and Stephen R. Quake, "The Biological Frontier of Physics," *Physics Today* 59 (May 2006): 38–43. © 2006, American Institute of Physics.

magnitude (Figure 4.6). This creates the exciting possibility that the molecules in our bodies can *spontaneously* convert different types of energy into one another. Molecules and small, nanoscale particles can have substantial fluctuations in energy as they take energy from the molecular storm (thermal energy), use it to convert, for example, chemical energy to electrical energy, and then release the energy again into the surrounding chaos. By contrast, smaller structures, such as atoms or nuclei, have binding energies that are too large to allow thermal energy fluctuations, unless the temperature (along with thermal energy) is extremely high (thousands or millions of degrees). At such high temperatures, molecules are unstable and the formation of complex structures needed for life is impossible. On the other hand, at scales much larger than a nanometer, mechanical and electrical energies are too high to be subject to thermal fluctuations. At this scale, everything becomes deterministic, and objects do not spontaneously change shape or assemble themselves—which are attributes needed for life.

Thus, the nanoscale is truly special. Only at the nanoscale is the thermal energy of the right magnitude to allow the formation of complex molecular structures and assist the spontaneous transformation of different energy forms (mechanical, electrical, chemical) into one another. Moreover, the conjunction of energy scales allows for the self-assembly, adaptability, and spontaneous motion needed to make a living being. The nanoscale is the only scale at which machines can work completely autonomously. To jump into action, nanoscale machines just need a little push. And this push is provided by thermal energy of the molecular storm.

But doesn't the molecular storm always lead to chaos, as suggested in the discussion of the second law of thermodynamics in Chapter 3? The answer would be yes if the molecular machines of living cells were just any old molecules—but they are not. They are clever little machines that can sift order out of chaos. How? Let's find out.

5
Maxwell's Demon and Feynman's Ratchet

Now let us suppose that . . . a vessel is divided into two portions, A and B, by a division in which there is a small hole, and that a being, who can see the individual molecules, opens and closes this hole, so as to allow only the swifter molecules to pass from A to B, and only the slower ones to pass from B to A. He will thus, without expenditure of work, raise the temperature of B and lower that of A, in contradiction to the second law of thermodynamics.

—JAMES CLERK MAXWELL, *THE THEORY OF HEAT*

The Moving Finger writes; and, having writ,
Moves on: nor all thy Piety nor Wit
Shall lure it back to cancel half a Line,
Nor all thy Tears wash out a Word of it.

—OMAR KAYYAM, 1048–1131

JAMES CLERK MAXWELL (1831–1879) LEFT BEHIND A distinguished scientific legacy. He unified electricity and magnetism, discovered electromagnetic waves and explained the nature of light, solved the riddle of Saturn's rings, developed modern color theory, laid the foundations for engineering control theory, and cofounded statistical mechanics. In addition to all this, he invented a demon.

The Scottish physicist's work on thermodynamics and statistical physics, *The Theory of Heat*, remains an example of clarity. Describing the second law of thermodynamics, Maxwell wrote: "One of the best established facts in thermodynamics is that it is impossible in a [closed] system . . . which permits neither change of volume nor passage of heat, and in which both the temperature and the pressure are everywhere the same, to produce any inequality of temperature or pressure without the expenditure of work. This is the second law of thermodynamics, and it is undoubtedly true as long as we can deal with bodies only in mass, and have no power of perceiving or handling the separate molecules of which they are made up."

This quote presents as clear a definition of the second law as we could hope for. The law forbids the creation of temperature or pressure differences in a uniform medium, unless work is expended to create the difference. Yet at the end of the quote, Maxwell included a caveat: As long as we deal with many molecules ("in mass") and have no way to look at them individually, the second law is true. But what if we *could* look at molecules individually? Maxwell continued: "But if we conceive a being whose faculties are so sharpened that he can follow every molecule in its course, such a being, whose attributes are still as essentially finite as our own, would be able to do what is presently impossible to us." Could such a being violate the second law?

Maxwell's Demon

The second law was invented to explain certain limitations of machines: It was well known that steam engines wasted a lot of energy—most of the energy supplied in the form of coal was wasted as heat and not turned into mechanical work. Physicists and engineers from the late 1700s to the mid-1800s were preoccupied with improving the efficiency of engines. The efficiency is the ratio of useful energy generated to the energy input in the form of fuel. For example, a car engine has an efficiency of about 25 percent. Only 25 percent of the gasoline we put into the tank is used for moving the car or running the electrical systems; the rest is lost as heat. Is there a limit to the efficiency of engines? Could a 100 percent efficient engine be made, at least in principle?

The new science of thermodynamics, and the understanding that heat was a form of energy, led Helmholtz to his universal law of energy conservation. Helmholtz convincingly demonstrated that motion and growth of a living being had to be powered by chemical fuel—food. In this sense, a living organism was similar to an engine. Like an engine, it converted a high-quality form of energy to both motion and heat. Again the question arose: Was there a limit to the efficiency of life's engines?

By the beginning of the twentieth century, it was becoming clear that the engines of life operated at the molecular scale. How can we understand such machines, and how does their operation relate to the macroscopic machines of our everyday experience?

Macroscopic machines—car engines, power plants, etc.—exploit *gradients*, that is, differences in temperature or pressure, to convert fuel into motion. This important observation was first made by a young French military engineer, Sadi Carnot (1796–1832), in 1824. Carnot laid the groundwork for modern thermodynamics when he realized that the efficiency of engines could not be increased to 100 percent, even in principle, but instead was limited by the temperature gradient they exploited. In other words, the hotter the fire, and the colder the surroundings, the more efficient a machine could become. When the inside of an engine approaches the temperature of the surroundings, no more work can be done by the engine, and the efficiency goes to zero. We then have the situation Maxwell described in his definition of the second law.

In living cells, temperatures and pressures are uniform—there is no combustion chamber or pressure reservoir. There are no temperature or pressure gradients. According to Carnot, no engine should be able to operate in our bodies. The second law of thermodynamics allows us to extract work from gradients, at the cost of creating waste heat and the leveling of the gradient. The result is equilibrium—a state of uniform temperature and pressure, a state from which no further work can be extracted. How can molecular machines extract work from the uniform-temperature environment of cells without violating the second law of thermodynamics?

When the second law was formulated in the nineteenth century, physicists were not certain if it was an incontrovertible law of nature. What was the basis of this law? Physicists like to employ many methods to get to the bottom of things: experiments, theoretical calculations, and gedanken-experiments. *Gedankenexperiment* is a German word, sometimes used in English, meaning "thought experiment." It is a hypothetical situation, which can only exist in thought and is a stress test for physics theories. The idea is either to create a paradox—a contradiction between different physical theories—or to see how far you can stretch a theory or an experimental result into a realm that is inaccessible to the real world. Famous thought experiments include Schrödinger's cat (which showed the absurdity of some interpretations of quantum mechanics); Galileo's deduction of constant acceleration during free fall, irrespective of the mass of the falling object (Galileo never threw objects off the Tower of Pisa—he inferred what would happen from rolling balls down inclines); Newton's cannonball (which showed that the motion of celestial objects is related to the falling of objects on Earth); and Maxwell's demon, the little creature who could transfer heat from cold to hot.

Maxwell's demon, which he devised in 1867, was a tiny hypothetical creature who controlled a little door separating two gas-filled chambers, which initially have the same average temperature (Figure 5.1). The job of the demon was to separate gas molecules into fast and slow molecules. If, for example, a fast molecule approached the demon's door from the right, the demon would let it through the door to the left chamber, but if it was slow, he would not. Conversely, he was happy to let slow molecules pass from the left to the right chamber, but not fast ones. Soon, the demon had sorted fast molecules into the left chamber, while leaving all the slow ones in the right chamber. Starting from a uniform-temperature system, the demon had created a temperature gradient—making one side cold

FIGURE 5.1. Maxwell's demon sorts fast-moving molecules into the left chamber and slow ones into the right chamber by controlling a small trap door. After a while, he succeeds in creating a temperature difference between the chambers without expending work, thus seemingly violating the second law.

and the other side hot (remember that the temperature of a gas is directly related to the speed of the gas molecules). This temperature gradient could now be used to do work if a little turbine were placed into the demon's door. The result would be to extract useful work from a uniform-temperature system, in clear contradiction of the second law. On the other hand, such a molecular Maxwell demon would be just what is needed to explain molecular machines! Are our cells full of molecular Maxwell demons? Does life, deep down, violate the second law?

Smoluchowski's Trap Door

Marian von Smoluchowski (1872–1917) spent a great deal of time thinking about the second law. An avid mountain climber, the Polish-Austrian physicist wrote prodigiously about everything, from the folding of mountains and erosion by glaciers to diffusion in colloids and heat transfer in liquids. He published his explanation of Brownian motion almost simultaneously with Albert Einstein, although he later admitted that Einstein's solution had the correct prefactor, and his didn't.

One of his many papers, "Experimentally Demonstrable Molecular Phenomena, Which Contradict Standard Thermodynamics," was published in 1912. In this paper, he discussed special states of matter where fluctuations—spontaneous deviations from the average value of some property (pressure or density, for example)—suddenly become very large. This happens for some gases close to a phase transformation. Smoluchowski asked, What if we could use these pressure fluctuations to push a one-way door? When the pressure gets high enough, the door will open; when it is too low, the door will remain closed. This would be an automated Maxwell demon: Smoluchowski's trap door would only let high-velocity (high-pressure) molecules through, while rejecting low-velocity molecules. According to Smoluchowski, several physicists of the age considered such hypothetical contraptions a serious objection to the second law. But Smoluchowski dismissed this idea. The problem, according to him, was in how strong the door was. If the door was very weak, or easy to open, it would be subject to thermal motion and would randomly open by itself—letting slow molecules through when it shouldn't, or letting fast molecules escape back into the slow pool. If the door could be made strong enough to avoid this problem, it might not open at all. And for values in between? The door would be unreliable. Sometimes it would not open when it should, and sometimes it would open when it shouldn't. Smoluchowski did not present a rigorous calculation in this paper, but he asserted that such a trap door would never work. The second law could not be violated using an automated trap door.

Toward the end of the paper, Smoluchowski emphatically pointed out that the second law *could* be violated—if we were willing to wait long enough. This is charmingly illustrated in physicist George Gamow's series of novels in which the hero, bank clerk Mr. Tompkins, learns about physics. In one book, Maxwell's demon makes Mr. Tompkins's highball boil, which prompts Mr. Tompkins's friend, the "professor," to excitedly exclaim how lucky they are: "In the billions of years to come, we will still, probably, be the only people who ever had the chance to observe this extraordinary phenomenon!" The second law is a statistical law—which states that *most of the time* (usually very close to "always"), systems tend toward their most probable state. However, if we are willing to wait long enough, strange things can happen by chance. I could win a million in roulette. Your cold coffee could spontaneously boil. But we'd be waiting a very, very long time for these things to happen.

How long? The waiting time depends on the probability, which in turn depends on the size of the system. In a large system, one visible with an optical microscope or larger, a violation of the second law will, for all practical purposes, never happen. However, a really small system (a single molecule, for example) can seemingly violate the second law relatively often. This was Maxwell's point when he invented the demon. Maxwell was not out to disprove the second law. He simply wanted to show that the second law *emerges* once we talk about a large number of molecules. It is a statistical law. This is why we usually do not apply the second law to single molecules.

Smoluchowski went on to point out that even though a small system may be able to violate the second law once, you cannot build, from such a small system, a device that can act like Maxwell's demon. The reason is that violating the second law happens randomly and not *repeatedly*. Consequently, the second law should be stated this way: You cannot *repeatedly* extract energy from a uniform heat bath.

But if it could happen once, why not repeatedly? Let's look at why not.

The Demon and the Reset Button

Maxwell's demon has been giving physicists a headache for a long time. Although Maxwell invented the little creature to show that the second law was only a statistical law, physicists later saw it as a serious assault on the law itself. In the interest of professional pride, the demon had to be exorcised.

Several solutions were proposed. One solution, proposed by the physicists Leo Szilard (1898–1964) and Leon Brillouin (1889–1969), posited that the very act of measuring the speeds of molecules would necessarily involve some energy, which would be dissipated. For example, the demon could try to measure the molecular speed with light pulses—but not all the light energy would be recoverable. The dissipated energy would lead to an increase in entropy larger than the decrease in entropy from the sorting of the molecules. However, more recently, computer scientists Rolf Landauer (1927–1999) and Charles Bennett (b. 1943) have pointed out that the measurement *could* be done without energy loss or entropy increase. Instead, the entropy increase would occur when the demon erases his short-term memory to make room for the next measurement. This solution to the demon conundrum created an intriguing connection between entropy and information storage, which is still a hot topic today.

Maxwell's demon illustrates why there is a connection between missing information and entropy. As discussed in Chapter 3, the reason the entropy of a gas at equilibrium is high is that it could be in many different microstates, which all would be compatible with the observed macrostate (pressure, temperature, and so on). But we do not know what the particular microstate of the gas is at each moment in time. Maxwell's demon, however, *would* know the microstate of the system, and thus, in some sense, reduce the entropy of the system just by having more information about the system (reducing the missing information). But in order to *repeatedly* learn more about the microstate of the gas (which changes all the time as gas molecules collide), the demon would need to erase old information to make space for the new, turning knowledge back into missing information. This erasure, according to Landauer and Bennett, comes with an energy cost. In other words, each time you erase information, you dissipate energy and increase entropy.

In the case of the demon, erasing information restores or "resets" the system to its original state, allowing a new measurement cycle to begin. But since this erasure leads to an increase in entropy, the demon could do his demonic deed only once without paying a price. He would not be able to do it "for free" repeatedly. The same applies to machines. To make a small machine perform repeated motions, a reset step is needed, as the machine needs to be returned to its original state before it can begin a new cycle. And it is this reset step that leads to an inevitable increase in entropy.

More recently, it has been pointed out that the entropy increase due to erasure of information assumes the second law, and therefore cannot be used to prove the second law. That would be a circular argument. Maxwell's demon may thus continue to haunt physicists' dreams. Whatever the correct answer to Maxwell's demonic challenge, my personal feeling is that Landauer and Bennett were on the right track. Clearly, such demons do not exist—otherwise, highballs would boil spontaneously, as in Gamow's tale of Mr. Tompkins. Any attempt to make a demon has failed (as evidenced by our inability to devise a perpetual-motion machine). Moreover, recent nanotechnology-based experiments have confirmed Landauer's conjecture that erasure of information creates entropy.

At the 2011 American Physical Society meeting in Dallas, physicists Yonggun Jun and John Bechhoefer of Simon Fraser University in Burnaby, Canada, reported on an experiment where they stored and erased information using a 200-nm plastic bead suspended in water. Following its motion with a microscope, they could calculate the heat associated when each bit of information was erased. A bit is a unit of information, contained in "yes" or " no," or "1" or "0." Landauer had postulated a minimum amount of dissipated energy when a bit of information is erased. This minimum is given by Boltzmann's entropy formula, times the temperature at which the bit is erased. Jun and Bechhoefer found that sometimes the heat released was less than Landauer's limit. This was due to water molecules occasionally helping the bead along. But this only happened at random times. In the long run—i.e., in the statistical limit—Landauer's limit stood. This work brilliantly confirmed Smoluchowski's and Maxwell's hunch that the second law *can* be violated, but not repeatedly and not predictably. It is a statistical law.

Reversibility

In 1918, Smoluchowski (posthumously) published a paper titled, "About the Concept of Chance and the Origin of the Laws of Probability in Physics." When this paper was published, the kinetic picture of matter, pioneered by Maxwell and Boltzmann, was already solidly accepted. Even the old critics, who had given Boltzmann such a hard time, had grudgingly accepted the existence of atoms and molecules. Probability now reigned supreme in theories of gases and liquids. The final blow to the anti-atomists was Perrin's experimental results, which completely confirmed the Einstein-Smoluchowski theory of Brownian motion.

Yet, just a few decades earlier, Boltzmann seemed to be fighting a losing battle. The kinetic theory was heavily criticized. One of the main objections was what Manfred Eigen would much later call Loschmidt's demon. Boltzmann had tried to show that the second law was a direct result of the motions of atoms—in other words, even in the case of just a few interacting atoms, the second law would hold. Boltzmann reasoned as follows: If molecules were initially in some low-entropy state, their collisions with other molecules would make their distribution of velocities more random, and increase entropy. However, the Austrian physicist Josef Loschmidt (1821–1895), a friend of Boltzmann's, pointed out that if that were true, what would happen if we reversed time? Loschmidt's demon was a powerful creature that could reverse time at will. In the realm of atoms, a collision obeys all known laws of physics, no matter if you play time forward or backward. Think of two billiard balls: Ignoring the player and the cue, concentrate on the moment when the two billiard balls collide. If you were to film this instant and play it to an audience forward or backward in time, it would be impossible to tell which is which. Simple elastic collisions, like collisions between molecules, are time reversible—they look the same run forward or backward. With this in mind, how could a time-*irreversible* law, like the second law, emerge from the *reversible* mechanics of molecules?

The answer to this conundrum was twofold. First, for molecules to move toward a more probable velocity distribution, they must be starting out with a less probable distribution. Thus, to see the second law in action, we have to assume that initially, the velocity distribution was improbable, and the entropy low. Then collisions shook things up, making the distribution more probable and increasing entropy. Thus, irreversibility came from the fact that the initial system was not at equilibrium. That is, it was not in a state of maximum entropy. This has consequences for the entire universe we live in: If there is such a thing as the arrow of time, which points from past to future, this arrow can only be there because the universe started in a very low-entropy state. Stars, galaxies, planets, and living beings have been feeding off the low entropy ever since.

The second part of the answer to how irreversibility can emerge from the reversible mechanics of particles is that the system has to be large enough—must contain enough molecules—so that collisions always mix things up. This is because in a large system, motions are generally uncorrelated, and molecular chaos reigns. If this is the case, what would happen in small systems?

In the late 1990s, a Los Alamos nuclear physicist, Christopher Jarzynski, derived an equality that electrified physics, especially the study of small molecular systems. Jarzynski's equality quantified how often small molecular systems violate the second law. As we have seen, small systems can violate the law at random times—and this is why, strictly speaking, we should not apply the second law to such small systems. Leaving this caveat aside, how often do molecules violate the law, and what would be the consequences?

The second law tells us any directed motion of a system will always encounter the resistance of friction. Friction is the result of many randomly moving molecules scavenging energy away from any nonrandom motion. Now let's imagine that a clever high school student has just learned about the conservation of energy. She devises a scheme for measuring the height of a mountain: Roll a ball down the mountain, starting the ball from rest, and measure its speed at the bottom. Then calculate the height. This calculation is an easy exercise, and I give problems like this to my introductory physics students. All you have to do is realize that according to energy conservation, the initial gravitational energy (ball on top of the mountain) has to equal the final kinetic energy when the ball reaches the bottom of the mountain. Gravitational potential energy is proportional to height, and thus equating the two energies, we can solve for the height (it comes out to $h = v^2 / (2g)$, where h is the height of the mountain, v is the speed of the ball at the bottom, and g is the acceleration of falling). But in practice, the measurement always falls short of the calculation. The kinetic energy of the ball at the foot of the mountain is *a little bit less than* the gravitational potential energy at the top of the mountain. This is because some of the energy is lost as heat, due to friction. Now the clever high school student decides to improve her accuracy by repeating the experiment a hundred times. Would that help? Not really; friction will always be there, and every single measurement will fall short.

Now let us imagine a similar experiment at the nanoscale. Shrunk to the nanoscale, our high school student repeats her experiment on a nano-size mountain. Most of the time, her measurements show the same trend as the macroscopic measurements: The speed is less than expected from the height of the mountain, because friction has taken its toll. But much to the nanoscale student's surprise, rarely and at completely random times, the speed of the ball is *more* than what is expected. The randomly moving atoms in the surroundings did not resist the motion of the ball, as one would expect, but actually pushed the ball along! When systems are small enough, there is a finite probability, though rare, that the atomic chaos surrounding the system actually adds energy to the system, rather than stealing energy.

Is there a way to combine measurements and find the height of the nanomountain? Yes, there is. Jarzynski's equality makes this possible, by averaging over an exponential function of the kinetic energy and not over the measured kinetic energies. Jarzynski showed that theoretically, you could obtain energy differences between two states (for example, top of the mountain and bottom of the mountain) from measurements in the presence of molecular chaos, and thus friction.

Experimental confirmation of this astonishing theorem did not have to wait long: Using laser tweezers, the biophysicists Carlos Bustamante and Jan Liphardt at University of California–Berkeley pulled on a single RNA molecule containing a loop. They wanted to know the energy difference between RNA molecules with the loop closed and with the loop open. But how could this be measured? Each time they pulled, they got a different answer. The surrounding water molecules created friction and made the measured energy difference between the open- and closed-loop states larger than the actual energy difference between the two states. One way to get close to the correct answer was to do the measurement very, very slowly. Going slow helps, because slow motion is associated with low kinetic energy, and if the kinetic energy is low, the surrounding atoms cannot steal as much. However, when they pulled on the loop with high speed, the measured energy difference was almost always higher than the values measured at slow speeds. Friction had taken its toll.

Sometimes, they saw the opposite, and the energy difference they measured was less than the minimum energy required to open the loop. This meant that the second law was occasionally violated. In these rare cases, randomly moving water molecules helped open the loop instead of resisting. Applying Jarzynski's formula, Bustamante and Liphardt averaged all their data, and the correct answer emerged. It was now experimentally confirmed: Nanoscale systems occasionally violate the second law of thermodynamics. At the molecular scale, entropy can sometimes spontaneously decrease (although, strictly speaking, entropy is not defined at this scale). When that happens, it is as if time has reversed.

Thus at the nanoscale, and for short times, Loschmidt's and Maxwell's demons *can* rouse from their slumbers and seemingly violate the second law. Could life's machines be Maxwell demons, creating order out of chaos by relying on the rare and unpredictable occasions when the second law is violated?

Perpetuum Mobile

To answer the question posed in the previous section, the answer is clearly no. All available evidence shows that life is not based, in any shape or form, on violating the second law. How do we know this? We know this since Lavoisier, Helmholtz, and many others determined that our bodies do not create energy, but rather waste energy. The efficiency of a human body (i.e., the amount of physical work obtained compared with the food energy intake) is about 20 percent. The rest (80 percent of food energy intake) is either directly turned into heat through friction or serves to maintain basic metabolic processes in our cells.

Helmholtz and Mayer had already realized that any violation of the first or second law by living organisms would mean that a perpetual-motion machine was possible—a machine that could either make energy out of nothing or operate at 100 percent efficiency. Both are impossible. The impossibility of a *perpetuum mobile* provided the main argument against the existence of a vital force. The argument went as follows: If there were a mysterious force that did not come from physical energy conversions, but was somehow inherent in life itself, such a force could add additional energy to a living system—energy that was not supplied as food. Helmholtz's obituary in the *Proceedings of the Royal Society* made a similar argument: "Helmholtz was led to the discussion of this subject [the conservation of energy] by reflections on the nature of the 'vital force.'" He had convinced himself that if it were true that living organisms could restrain or liberate the action of chemical or physical forces, perpetual motion would be realized." A life force could create energy out of nothing—a prospect that was clearly absurd. Even in the eighteenth century, most physicists firmly believed that the creation of a perpetual-motion machine was—in principle—impossible.

The impossibility of a vital force can be directly related to the impossibility of realizing a Maxwellian demon. This is the argument: Why did eighteenth-century biologists postulate a vital force? Because they were looking for something that could explain the "directed activity" seen in living organisms. It was a way to explain "purpose," to explain the apparent "intelligence" that seemed to operate in humble cells and microbes. This directed activity had to ultimately stem from the directed activity of the molecules that made up the organism. Who could direct this activity? Only a molecular Maxwell demon, who, by using his intelligence, injects purpose into the otherwise senseless motions of molecules.

But a Maxwell demon is impossible, because a perpetual-motion machine is impossible. The connection between information and entropy, made clear by the thought experiment of the demon, shows that intelligence, purpose, or vital forces can play no role at the molecular scale if the statistical second law of thermodynamics is supposed to hold. Before Maxwell's demon was conceived, Helmholtz and Mayer had already realized that vital forces had no place in the play of molecules, not even molecules that live in our cells.

Feynman's Ratchet

Our discussion leaves us empty-handed: To make life work, we need something *like* a Maxwell demon—something that can create directed activity out of chaos. Yet, a Maxwell demon is impossible. The existence of such a being or object would lead to the creation of perpetual-motion machines and violate the second law. The fact that the second law can sometimes be seemingly violated does not help much, either, because it can only be violated by single molecules, at random, and not repeatedly.

Strictly speaking, the second law is a statistical law, that is, the result of averaging over long times or many molecules. It is therefore not truly violated by rare events that seem to run counter to it. *On average*, nothing, not even molecules, violates the second law. And you cannot make a functioning machine out of a molecule that only works occasionally and usually moves in the wrong direction.

Marian Smoluchowski devised another simplified version of Maxwell's demon, which is very relevant to the mystery of molecular machines. Looking at Maxwell's demon, he wondered if it would be possible to devise a tiny machine that could extract work from the random motions of molecules in a uniform-temperature environment. There is energy contained in the random thermal motion of atoms, the molecular storm. But how could we harness this energy? Smoluchowski's machine consisted of a wire with a ratchet attached at the other end. Years later, physicist and Nobel laureate Richard Feynman read Smoluchowski's work and devised a particularly fruitful incarnation of Smoluchowski's machine: *Feynman's ratchet*.

What kind of molecular device could channel random molecular motion into oriented activity? Such a device would need to allow certain directions of motion, while rejecting others. A ratchet, that is, a wheel with asymmetric teeth blocked by a spring-loaded pawl, could do the job (Figure 5.2). Old-fashioned watches have ratchets in their windup mechanisms, as do pulleys. As the son of a watchmaker, I have seen tiny ratchets in windup wristwatches many times. The ratchet allows us to wind up our watch but not let it unwind. It allows easy motion in one direction, but blocks motion in the opposite direction. Maybe, nature has made molecular-size ratchets that allow favorable pushes from the molecular storm in one direction,

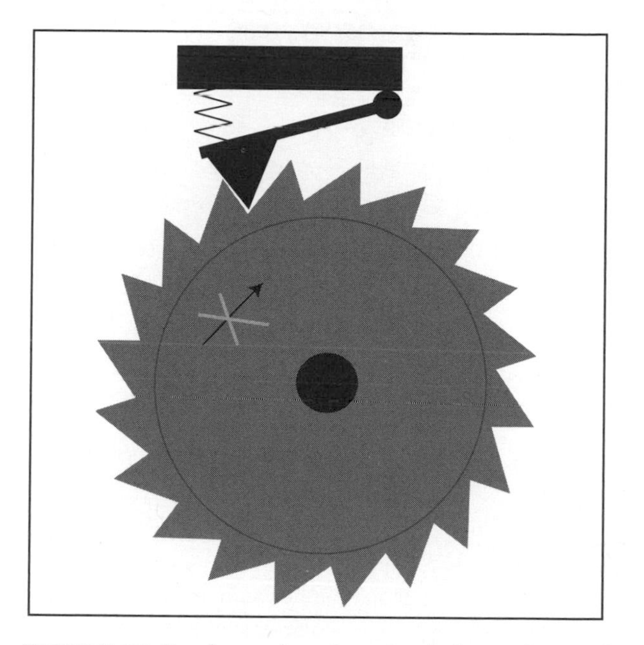

FIGURE 5.2. Ratchet and pawl—a simple device designed to block motion in one direction (clockwise in this case), but allow it in the other direction (counterclockwise).

while rejecting unfavorable pushes from the opposite direction. This surely would be a nice way to harvest energy from random motion.

I am sure you already sense that there is something fishy about this suggestion. After all, every other attempt to make machines, doors, or demons that could extract useful work from the molecular storm has failed. But surely, this one looks promising: You just need to adjust the spring of the pawl to the right stiffness, and the wheel should block backward motion, while easily sliding in the forward direction. What could go wrong with this idea?

The Ratchet Fails

Alas, Feynman showed that this hypothetical machine was also impossible. To get a molecular-size ratchet and pawl to work, the pawl needs to be spring-loaded so that it moves up and down, notch after notch. To allow the ratchet to rotate, the pawl must be pushed up to the height of one of

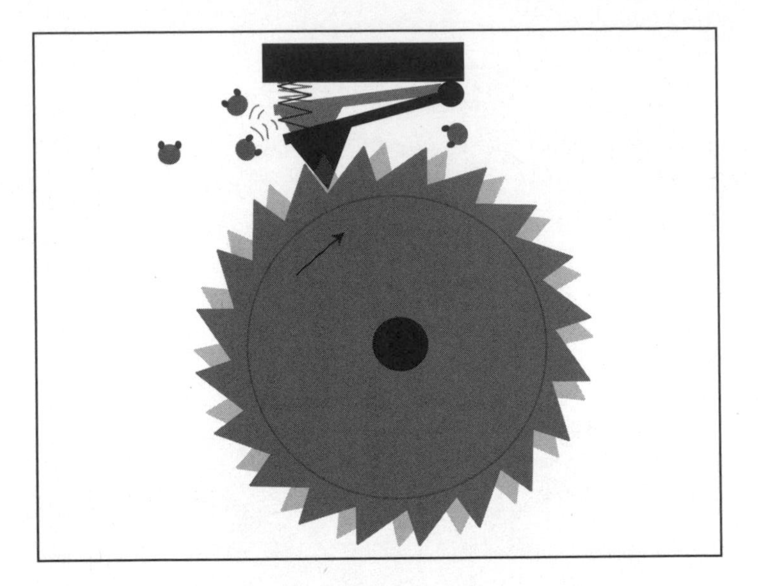

FIGURE 5.3. A molecular-size ratchet and pawl would need to have a weak enough spring to allow collisions with molecules to turn it. But such a weak spring would also allow the water molecules to randomly open the pawl, allowing the ratchet to slip backward.

the teeth. The backward step is restricted because a much larger force is needed to push the pawl up the steep edge of the tooth rather than the gentle incline on the other side. The energy or work is the same in either case, because work equals force times distance. Pushing the pawl up the gentle incline takes less force, but more distance, while pushing it up the steep edge takes more force and less distance.

For the ratchet-and-pawl machine to extract energy from the molecular storm, it has to be easy to push the pawl over one of the teeth of the ratchet. The pawl spring must be very weak to allow the ratchet to move at all. Otherwise, a few water molecules hitting the ratchet would not be strong enough to force the pawl over one of the teeth. Just like the ratchet wheel, the pawl is continuously bombarded by water molecules. Its weak spring allows the pawl to bounce up and down randomly, opening from time to time, allowing the ratchet to slip backward, as shown in Figure 5.3. Worse, because the spring is most relaxed when the pawl is at the lowest point between two teeth, the pawl spends most of its time touching the steep edge of one of the teeth. When an unfavorable hit pushes the ratchet backward just as the pawl has opened, it does not need to go far to end up on the incline of the next tooth, and the spring will push the pawl down the incline—rotating the ratchet backward! Feynman calculated the probabilities of the ratchet's moving forward and backward and found them to always be the same. The ratchet will move, bobbing back and forth, but it will not make any net headway.

Any simple device, when stuck in an isolated, uniform-temperature bath can only move randomly—no matter how ingeniously designed. The impossibility of making simple, passive machines that can extract oriented work from random thermal energy, be it Smoluchowski's trap door or Feynman's ratchet, is a powerful illustration of the second law of thermodynamics. Work cannot be repeatedly extracted from an isolated reservoir at uniform temperature. If it were possible to make machines that could do this, our energy problems would be solved: Such machines would convert heat in our environment back into ordered mechanical energy. Imagine placing such a contraption into your backyard. It would make the air in the backyard colder and turn the extracted heat into electricity. That would be wonderful, but alas, nobody has been able to violate the second law of thermodynamics. At least nobody larger than ten nanometers. And even at ten nanometers, it only happens randomly and rarely.

So, we seem to have hit a snag: We know our cells are full of tiny machines, and we know they must be molecular in size—but we have not yet explained how they work. Does life have the mysterious power to defy the second law of thermodynamics—a law that rules supreme in the inanimate universe? Or are we missing a crucial ingredient that allows life's molecular machinery to use the molecular storm without violating this all-powerful physical law?

6

The Mystery of Life

What is the characteristic feature of life? When is a piece of matter said to be alive? When it goes on "doing something," moving, exchanging material with its environment, and so forth, and that for much longer period than we would expect an inanimate piece of matter to "keep going" under similar circumstances.

—ERWIN SCHRÖDINGER, *WHAT IS LIFE*?

For molecules, moving deterministically is like trying to walk in a hurricane: the forces propelling a particle along the desired path are puny in comparison to the random forces exerted by the environment. Yet cells thrive. They ferry materials, they pump ions, they build proteins, they move from here to there. They make order out of anarchy.

—R. DEAN ASTUMIAN, "MAKING MOLECULES INTO MACHINES"

Erwin Schrödinger quotation courtesy of Cambridge University Press. R. Dean Astumian epigraph quote courtesy of R. Dean Asumian; from R. Dean Asumian, "Making Molecules into Motors," *Scientific American* (July 2001).

MOLECULAR MACHINES READ AND TRANSLATE DNA; MAKE new machines; operate the processes that makes cells reproduce, transport nutrients, and expel wastes; and help the cell change shape and move about. These tiny machines are the basis of life. But how do they work?

So far, our attempts to explain how molecular machines can do useful work in the midst of the molecular storm have been foiled by the second law of thermodynamics. No machine, however ingeniously designed, can directly convert the random thermal energy into "oriented, coherent activity," in the words of Jacques Monod. What are we missing?

Thou Shalt Not Violate the Second Law

In a September 2000 spoof in the satirical newspaper *The Onion*, concerned voters demanded that their legislators repeal the second law of thermodynamics: "'Why can't disorder decrease over time instead of everything decaying?' . . . 'Is that too much to ask? This is our children's future we're talking about!'" If the second law were a dogma imposed by nefarious physicists, Congress might tell us to get rid of it. But we are not that powerful (conspiracy theories about evil scientists notwithstanding). The second law is an *inescapable* (macroscopic) consequence of the randomizing power of the *inescapable* (microscopic) molecular storm.

As explained by physicists Dean Astumian and Peter Hänggi in their 2002 *Physics Today* article on Brownian motors, a typical molecular motor uses about 100 to 1,000 ATP molecules per second. This translates into performing work at a rate of only 10^{-16} watts (1 watt = 1 joule/second). It would take 10^{21} molecular motors to generate as much power as a typical car engine—an unimaginably large number. Yet, this number of molecular machines barely fills a teaspoon—a teaspoon that could generate 130 horsepower! The power density of molecular machines (the power generated per volume the machine occupies) is very large, about 10^{8} watts per cubic meter. A car engine has a power density that is a thousand times smaller. These little machines are extremely efficient: As mentioned earlier, our entire bodies, which are based on these machines, operate on a power "budget" of just 100 watts, the same as a (large) incandescent light bulb.

If that were not astonishing enough, consider the world these machines inhabit. At the nanoscale, nothing can escape the molecular storm. As Astumian and Hänggi point out, every molecular machine in our bodies is hit by a fast-moving water molecule about every 10^{-13} seconds. Each collision delivers on average 4.3×10^{-21} joules of energy (the energy is determined by the product of Boltzmann's constant and body temperature measured in degree Kelvin). This translates into an average power input of more than 10^{-8} watts. Remember that a molecular machine generates only about 10^{-16} watts in power. Thus, the power input from the random pounding of water molecules is a hundred million times larger than the power output of our machine! To put this into perspective, compare this to a car in a windstorm. How powerful would a storm need to be to transfer a hundred million times more power to a car than the car's engine can generate? Taking the air resistance of a car into account, the storm would need a wind speed of an astounding seventy thousand miles per hour to have the same effect that the molecular storm has on our hapless molecular machines! I doubt even a Hummer would get very far against a storm like that.

A Molecular Hummer

An important difference between a macroscopic storm and our molecular storm is that the molecular storm has no preferred direction. In other words, every collision with a water molecule comes from a random direction. Storms of our everyday experience blow in a more or less constant direction and can perform useful work (for example, move a sailboat or a windmill). At the nanoscale, not only is the molecular storm an overwhelming force, but it is also completely random. It is difficult to imagine how a molecular storm can serve as an energy source for useful work.

What to do? We can think of two possibilities. We could build an incredibly robust machine, a kind of molecular super-Hummer, which could make small, measured steps while resisting the thermal chaos surrounding it. Alternatively, we could make a "floppy" machine that works according to the motto "If you can't beat 'em, join 'em." That is, make a machine that could "harvest" favorable pushes from the random hits it receives. Since the latter was our approach in Chapter 5 and has so far failed miserably, it may be tempting to go with the first option. But the super-Hummer option would clearly not work, either. Such a molecular machine would need to be made exceedingly "stiff" to withstand the molecular storm. If it were that stiff, how could it move at all? As soon as it wanted to move, it would have to weaken its springs. But if the springs were weakened, the machine would be pushed around in the storm just as before. Moreover, where does motion come from in the first place? As we have seen, the only motion existing at the molecular scale *is* the molecular storm. How could a molecular machine move, while resisting this motion?

Let's stick with the notion that our machine must somehow tame the molecular storm, regardless of whether the machine is robust or floppy. To understand how molecular machines tame the chaos of thermal motion, we first need to learn what these machines actually look like. Obviously, they are not truly the ratchets or car engines we used as convenient analogies. To describe molecular machines, we need to learn a little bit of biochemistry.

Squeeze to Fit

Molecular machines must change their shapes to create locomotion. How is this possible? A good starting point is to look at chemical reactions: In chemical reactions, molecules must change shape to combine in novel ways. These changes in shape are driven by a reduction in free energy.

If this were all there were to driving chemical reactions, they would happen in a snap. Yet, many chemical reactions take time. Your car does not turn into a pile of rust overnight, and your milk does not curdle as soon as you open the carton. The speed of chemical reactions is explained by the concept of *activation energy*. During any chemical reaction, there is an awkward moment when molecules no longer have their original shapes, but neither do they have their final shapes. They form an intermediate state between their initial state (reactants) and their final state (products). This intermediate state, called the *transition state*, tends to be uncomfortable for the molecules.

Figure 6.1 illustrates the transition state with the closing of a cardboard box. One way to close a cardboard box is to overlap each of the four flaps with a neighboring flap. When I try to do this, there is always a tricky moment when I am trying to push down all four flaps at once—one always pops out. But once I've succeeded, the box is perfectly happy and shut.

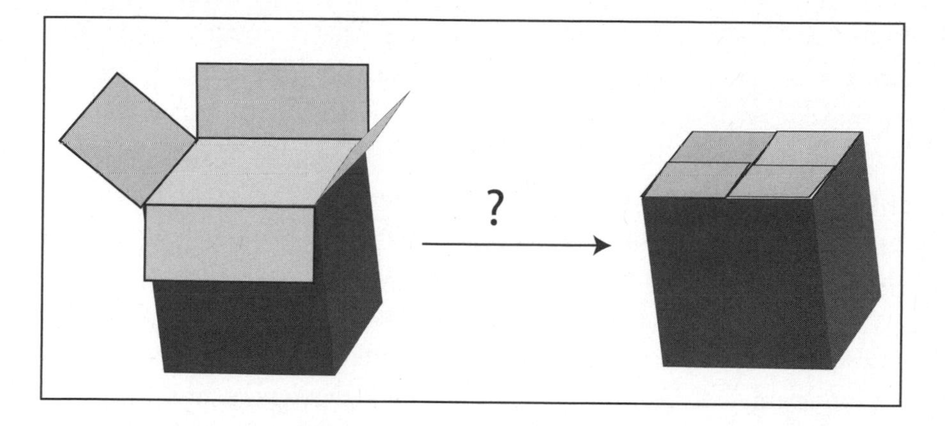

FIGURE 6.1. Folding a cardboard box to close it. Both the open and the closed box are stable states, but to get from one to the other, we need to push all four flaps together at once, an often quite unstable and awkward state of the system.

Thus we have two comfortable (low-energy) states. Initial state: box (and flaps) open; final state: box closed. To get from one state to the other, we have to cross an intermediate uncomfortable, high-energy state (shown as a question mark in the figure).

The same holds true for chemical reactions. As reactants turn into products, chemical bonds become strained for a moment before the molecules relax into their final shapes. This strained state has higher energy than either the original or the final state of the transformation. In Figure 6.2, two molecules are not quite compatible, but are "trying" to combine and form a new compound molecule with lower energy. The two molecules can lower their energy if they combine, but first they need to obtain enough energy to pass over the transition state. In the cardboard box example, I needed to push the flaps down with enough force so they snapped into place. In other words, I needed to supply sufficient energy to get past the awkward arrangement of the four flaps pushing against each other. In chemistry, the energy required to cross over the transition state is called the *activation energy*. It is the minimum amount of energy required to activate a chemical reaction.

Where do molecules obtain the needed activation energy? From the molecular storm! The impetus needed to make it across the transition state is provided by small, fast molecules (typically water molecules) fortuitously colliding with the reacting molecules to give them the right push. If lucky, the push causes the molecules to snap into their new shapes. Of course, not every colliding water molecule will have enough energy or hit the reacting molecules in the right way. Chemical reactions take time—we have to wait for the right push to come along, and the higher the activation energy needed, the more time it takes, as higher-energy collisions are much less frequent than low-energy collisions.

What determines the height of the activation barrier? In a nutshell, it depends on how uncomfortable the molecules become as they transform from the initial to the final state. If I try to change a molecule by rearranging several of its bonds, it will take a lot of bending and twisting, and activation energy will be high. On the other hand, if a tiny rearrangement of one atom will do, activation energy may be low.

Many important chemical reactions in cells require enormously large activation energy. These transformations proceed at such a glacial pace we'd probably be dead before our cells cycled through them even once. To solve this problem, nature has invented a neat trick. Let's imagine a helper

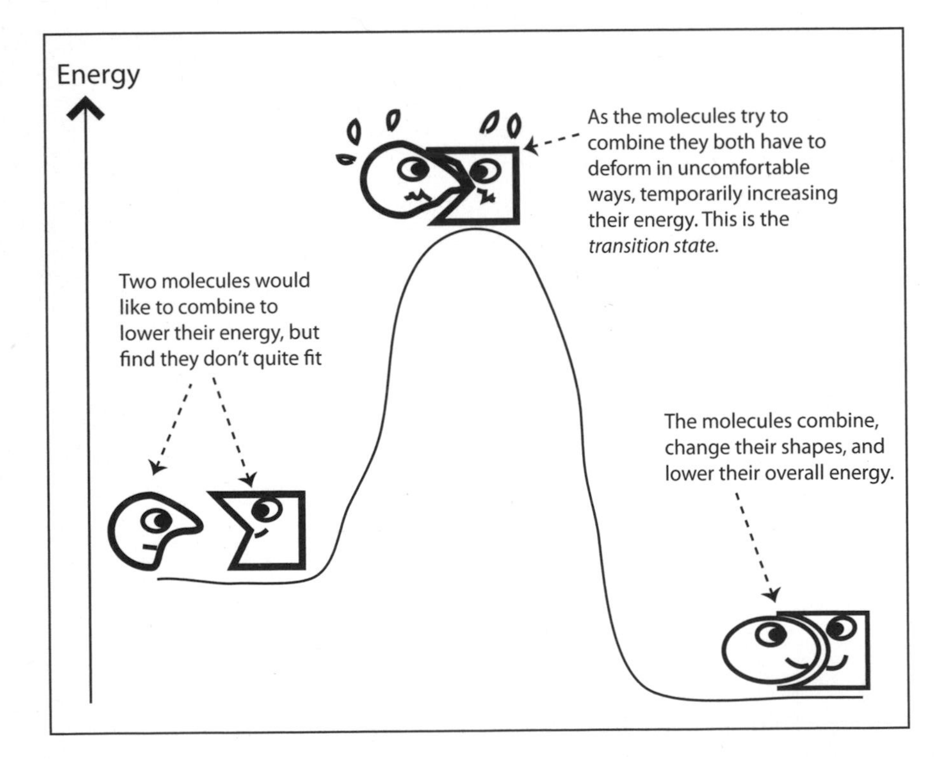

FIGURE 6.2. Chemical reactions proceed via "uncomfortable" transition states.

molecule whose job is to make the transition state more comfortable and therefore lower the activation energy (Figure 6.3). If the activation energy is lowered, the reaction rate increases. Such helpful molecules are called enzymes, and their involvement speeds up reactions by many orders of magnitude. Enzymes are biological catalysts, comparable to your car's catalytic converter, which speeds up the oxidation of harmful exhaust gases.

Enzymes have a pocket that is custom-made to fit the substrate (the reactant molecules). Initially, when the substrate binds into the enzyme's pocket, it is not a perfect fit, but as the enzyme and the substrate slightly change shape and begin to fit more comfortably, the enzyme pushes the substrate toward the transition state. Once the substrate is in the transition state, the enzyme is happy (at lower energy), and the total energy of the system (the energy of the transition state plus the energy of the enzyme) is reduced (Figure 6.3). Because the transition state still has higher energy than the final, product state, the transition state has no choice but

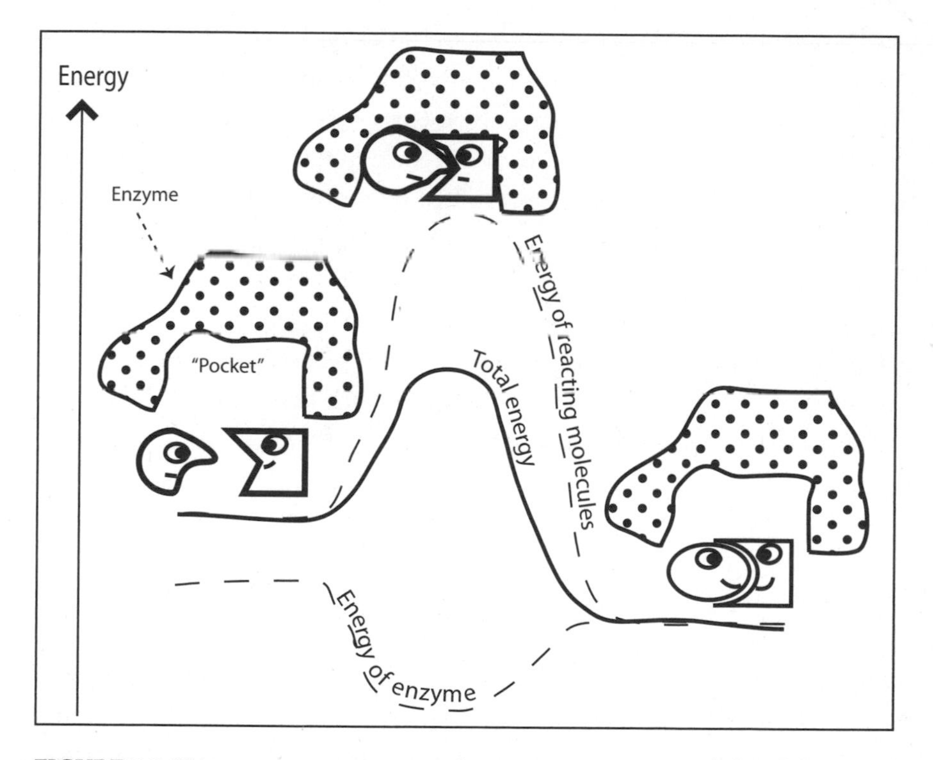

FIGURE 6.3. How an enzyme accommodates a transition state and therefore lowers the activation energy of a chemical reaction.

to convert to the product. The product is uncomfortable in the enzyme's pocket and is released. The enzyme snaps back to its old shape and is ready to take on a new substrate for conversion.

To illustrate the remarkable abilities of enzymes as catalysts, consider an enzyme called phosphoglucomutase. This enzyme converts an indigestible kind of sugar into a more palatable one. By speeding up the conversion process by a factor of one trillion, one phosphoglucomutase molecule can convert a hundred sugar molecules per second. Without this enzyme, it would take three hundred years to convert just one sugar molecule!

Enzymes are large molecules made of proteins—long chains of amino acids that fold into complicated shapes, as described in Chapter 4. Proteins are like a box of Legos: By changing the sequence of amino acids, almost any shape of protein can be constructed. Using the trial and error of evolution, myriads of enzymes have evolved, each custom-made to catalyze a particular reaction.

Molecular machines can be considered a special type of enzyme. As will be described later, they have evolved from enzymes. Enzymes and molecular machines share many attributes. In some sense, enzymes *are* machines: They bind, transform, and release molecules by virtue of their specific structure, but also as a result of being jostled by the molecular storm. A molecule does not know how it will fit into the binding pocket of the enzyme. Instead, the constant bombardment by water molecules (about every 10^{-13} seconds) rapidly rotates and deforms the molecule. After a millisecond, by chance, the molecule is pounded into the right shape and orientation to form a complex with the enzyme. If you are amazed that the molecule is able to find the right orientation and shape, consider that in a millisecond, an average molecule undergoes ten billion collisions with water molecules.

Being in Control

Enzymes are amazing molecules, but they are not what we would call alive. First of all, they do not move. They deform a little bit when binding a substrate molecule, but this seems hardly sufficient to operate a molecular machine.

In the early days of biochemistry, biologists envisioned the cell as a pouch containing enzymes that catalyzed various reactions. The cell was thought of as nothing more than a tiny reaction vessel. It soon became clear, however, that cells would not survive long if enzymes were given free rein to make and break down molecules. Cells have to respond to their environment. Cells have to be able to make decisions.

Computers can make decisions because computer programs contain logical commands such as "IF . . . THEN. . . ." In a cell, logical decisions are implemented on a molecular scale. Part of this cellular software is, of course, DNA, which contains the blueprint for how to make different proteins and how to guide a body through its development. But DNA does not control the day-to-day operation of the cell. This is done by enzymes and molecular machines. So the real question is, how do enzymes compute?

This mystery was solved in the early 1960s by Jean-Pierre Changeux (b. 1936), a young graduate student in Jacques Monod's laboratory at the Pasteur Institute in Paris. Studying an enzyme called L-threonine deaminase, which broke down its substrate, the amino acid threonine, he found that the enzyme's activity was inhibited in the presence of the molecule isoleucine, another amino acid. Inhibition of enzymes had been seen before and was thought to occur when the enzyme was bound to a molecule other than its substrate. This process of an unwanted houseguest's clogging up the binding pocket of an enzyme is called competitive inhibition. A familiar example is carbon monoxide poisoning: This gas will fit so strongly into the oxygen-binding pocket of hemoglobin (the oxygen-carrying proteins in our red blood cells) that it cannot be released again. In the competition between oxygen and carbon monoxide, carbon monoxide is the clear winner. Once the hemoglobin in a victim's blood becomes gummed up with carbon monoxide, oxygen transport is shut down, and the hapless victim suffocates.

Changeux's deaminase did not follow the model of competitive inhibition, however. When he tried to shut down the isoleucine binding site on the enzyme, the enzyme continued its enzymatic activity uninhibited. If the isoleucine binding site and the catalytic binding pocket had been one and the same—as in the case of hemoglobin—the enzyme's ability to bind the threonine substrate should have been impeded. Changeux concluded that the enzyme had *two* binding pockets: one for threonine (the enzyme's substrate), and one for the isoleucine, which regulated the enzyme's activity.

In his autobiographical paper "Allosteric Receptors: From Electric Organ to Cognition," Changeux recalls that he first presented his startling findings at the Cold Spring Harbor Symposium on Quantitative Biology in 1961. Here, he told his fellow scientists that "the interaction between these two sites was indirect and transmitted by a conformational change of the protein molecule." In other words, binding the isoleucine caused a large change in the shape of the enzyme, closing down the binding pocket for the substrate, and rendering the enzyme unable to process threonine molecules (Figure 6.4). The isoleucine acted as a control molecule—it was as if in a computer program, there were a line: "IF isoleucine is present THEN do not process threonine."

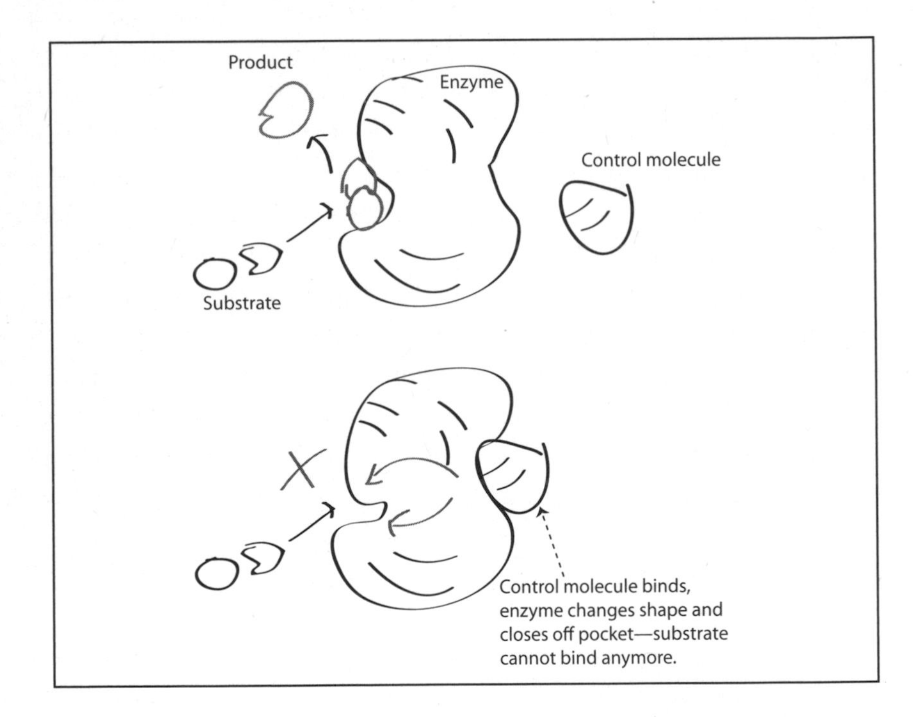

FIGURE 6.4. Top: An allosteric enzyme combines two smaller molecules (its substrate) into one larger molecule by binding them to a suitable pocket in its structure (active site). A control molecule (right) approaches the enzyme. Bottom: The control molecule has attached itself to the enzyme's control site. This has caused the enzyme to change shape (conformation) and close off the active site. The enzyme halts its catalytic activity as long as the control molecule remains bound.

Much enzymatic activity in living cells is controlled by the binding and release of such control molecules. Binding a control molecule changes an enzyme's shape, with part of the enzyme moving relative to other parts. This change in shape can increase or decrease the enzyme's catalyzing activity by opening or closing the pocket that binds the enzyme's substrate. The control molecules can also completely shut down the enzyme. This ability of a control molecule to change the enzyme's shape and activity was named *allostery* by Changeux's thesis advisors, Jacques Monod and François Jacob.

Once you have allostery, you can implement many useful schemes to control chemical reactions in a cell. Imagine that an enzyme creates a product, which also serves as the enzyme's control molecule. This would allow a product to regulate its own production. Such circular schemes are known as feedback loops. A familiar feedback loop is the feedback between an audio speaker and a guitar pickup—when a rock guitarist holds her guitar against the speaker, the sound from the speaker causes the guitar strings to vibrate. This is picked up by the pickup, amplified, and fed back to the speaker, which now makes an even louder sound, vibrating the guitar strings even more. The result is an increasingly loud shriek, only limited by the power of the amplifier. This kind of feedback is known as *positive feedback*—a feedback where the product (sound, in our example) enhances the production of more product. Positive feedback also exists in cells—some enzymes speed up production in the presence of product. The result is a rapid, explosive increase in product (until the reactants run out). This can be useful if the cell needs to produce a chemical very quickly in response to an external stimulus.

More common is the opposite case: *negative feedback*. An example is Changeux's L-threonine deaminase. This enzyme is part of a number of enzymes that work together to make isoleucine, starting from threonine. But isoleucine was also the control molecule that inhibited the activity of L-threonine deaminase. Thus, the product inhibits its own production. As a result, the enzymes make just enough product until the product molecules shut down further production. This is a nice way to control the maximum amount of a product molecule in a cell.

There are more complicated schemes that involve vast networks of interacting enzymes. The product of one enzyme may act as the control

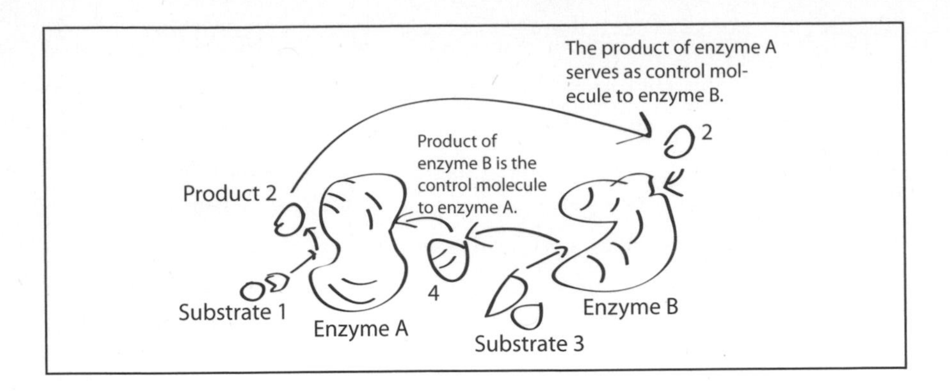

FIGURE 6.5. A molecular feedback loop. The product 2 of enzyme A serves as a control molecule for enzyme B, while the product 4 of enzyme B serves as a control molecule for enzyme A. Depending on the control molecule's influence on the enzymes (inhibition or enhancement), several programs can be implemented this way. In addition, substrates 1 and 3 may be the products of other enzymes, which themselves are controlled by control molecules. This way, more and more complex molecular programs can be constructed.

molecule for another enzyme, either enhancing or inhibiting its activity. The product of this second enzyme may again control the first enzyme, forming a two-enzyme feedback loop as shown in Figure 6.5, or the product may influence a third enzyme, which influences a fourth, and so on. Complicated schemes of feedback loops and mutual enhancement or inhibition provide the computing power that makes living cells seem intelligent.

Rambling Across the Energy Landscape

Allosteric enzymes bring us closer to our understanding of molecular machines. A large change in the shape of an enzyme, when it binds to a substrate or control molecule, can be seen as a type of motion. We could imagine, for example, that this change in shape creates forward motion when we place an allosteric enzyme on a molecular track: Each time the enzyme changes shape, it pushes against the track and propels itself forward (Figure 6.6). Examples of molecular tracks include fibers and filaments, typically made of long strands of proteins. One example is actin, which we already encountered in Chapter 4. Another example is DNA—and indeed there are specialized machines that move along DNA like lo-

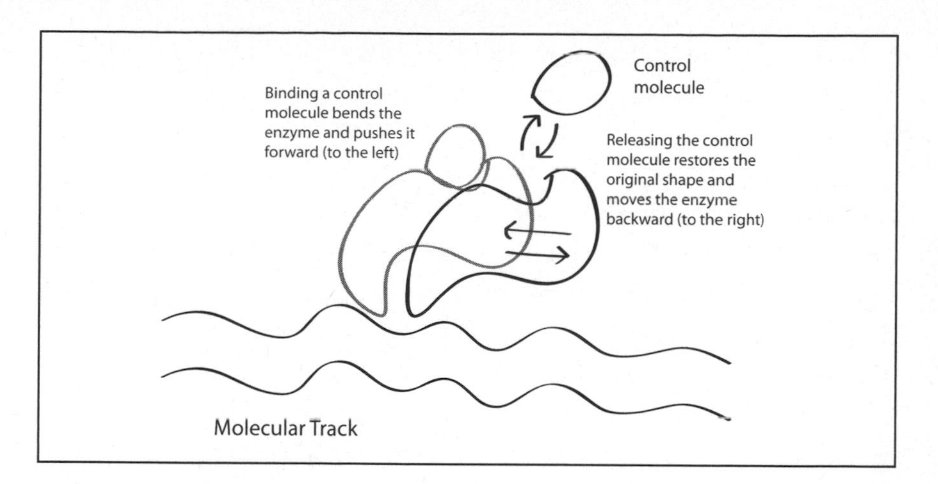

FIGURE 6.6. An allosteric enzyme bound to a molecular track. As the control molecule binds, the enzyme changes shape and pushes the enzyme forward along the track. However, the unbinding of the control molecule pushes the enzyme back to where it came from. To move forward, it would need to detach temporarily.

comotives on a train track. We will meet some of these machines in the next chapter.

A molecular machine, however, is not simply an allosteric enzyme bound to a track. The problem with this idea is that an allosteric enzyme exists in exactly two states: It is in shape A if the control molecule is attached, and shape B when the control molecule is released. The enzyme may push itself forward when binding a control molecule, but would revert to its old shape and move backward once the control molecule is released. Just as with our ratchet in Chapter 5, the molecule would dither back and forth, but make no headway. To move forward, the machine needs to loosen its grip on the track during the back step and only attach during the forward step. But if the machine loses its grip on the track, the molecular storm will sweep it away.

Sisyphus at the Nanoscale

It seems we have returned to our original problem and must therefore add another step: The machine needs to reset itself to its original shape without creating backward motion. The machine needs a reset button, just like

Maxwell's demon. Remember, Maxwell's demon could only work if he received external energy to erase his memory, that is, if he received energy to reset.

For a machine moving along a molecular track, the reset step would consist of detaching from a track temporarily, releasing the control molecule, and resetting the machine's shape. Then the machine would rebind to the track, bind to a control molecule, and push forward against the track—repeating the cycle. This idea leaves us with a few questions: Why doesn't the motion of the machine become randomized during the reset step? That is, why does the detached machine not get swept away by the molecular storm? And how does the presence of a reset step help explain how molecular machines avoid violating the second law?

To try to answer these questions, let us turn to Sisyphus, the mythical character condemned by the gods (for a certain divinity-angering faux pas) to push a large, heavy boulder up a steep hill for eternity. If he could get the boulder to the top of the hill, his penance would be fulfilled. Alas, his boulder was cursed: Every time he approached the hilltop, the boulder slipped from his hands and rolled back down. What makes pushing a boulder so difficult? When we push something up an incline, we have to use enough force to overcome friction and gravity. As we apply this force and move the boulder over a distance, we perform work (in physics, work is the product of force and distance), and doing work requires energy.

Imagine if Sisyphus and his boulder were denizens of the nanoworld. If the boulder were nanosize, it would be continuously pushed around by the molecular storm. Ignoring sideways motion, the boulder would be randomly pushed up and down the hill (more often down than up, because up requires more energy). Sisyphus could adopt a very energy-efficient strategy to get the boulder to move uphill: simply wait for the boulder to move uphill by itself, randomly pushed by the molecular storm. But this is no good—the boulder is more likely to be pushed downhill than uphill.

What if Sisyphus steps in the way of the descending boulder? He could block the boulder's downward motion, while allowing it to move uphill (Figure 6.7). He could repeat this all the way up the hill, as long as he quickly steps behind the boulder each time it makes a jump uphill. Now we are getting somewhere: The boulder is propelled by the molecular storm, but this random motion is rectified by Sisyphus's blocking its motion in the undesired direction. This is an extremely efficient way to move the boulder—a nano-Sisyphus would not have to push the boulder, and the boulder would move itself, driven by the molecular storm. Does this mean Sisyphus doesn't need energy to move the boulder? No! Sisyphus still needs to move up the slope, step-by-step. These steps require energy.

How is this different from the ratchet in Chapter 5? The ratchet did not have a reset step and did not require energy. The ratchet was a passive device, which, by itself, was supposed to rectify thermal motion. But Sisyphus is different: He rectifies thermal motion at the expense of using a source of energy to reset the system (stepping up behind the boulder).

Let us take this analogy further. When nano-Sisyphus steps up the hill, he converts some source of free energy into kinetic energy. Then he stops. His kinetic energy is turned into heat (his sandals are hitting the ground). The act of stepping up the hill degrades free energy into heat. Sisyphus and the ground are getting a bit warmer. Now, whenever the boulder tries to move downward, it bounces off Sisyphus (who is warmer), and some energy is transferred to the boulder (which is colder). In the end, Sisyphus and the boulder end up at the same temperature (thermal equilibrium), but as a result, some of Sisyphus's free energy ends up moving the boulder uphill, even though Sisyphus is not actually pushing the boulder!

Moreover, this free energy is ultimately degraded into heat, as demanded by the second law of thermodynamics. We have found a way to

FIGURE 6.7. Left: A macroscopic Sisyphus is condemned to sweat as he haplessly pushes the boulder uphill. Middle: A nanosize Sisyphus wonders how he can utilize the random thermal motion of the nanoboulder. Right: Nano-Sisyphus comes up with an idea: Step behind the boulder as it descends, but let it move freely when it happens to go uphill.

use the molecular storm to move the boulder without violating the second law. Without the chaotic motion of the molecular storm, the boulder would not move at all, and Sisyphus would be condemned to push forever. Yet, at the nanoscale, chaos can be turned into order, as long as we have a supply of free energy to periodically reset our machine.

The reset step, which we have found to be necessary for a molecular machine to work, can take many forms. This step is an example of a so-called *irreversible* step, because it degrades free energy into heat, and this heat cannot be turned back into free (usable) energy. Through this irreversible step, our molecular machine can extract energy from the molecular storm without violating the second law.

In hindsight, this makes sense: A reversible machine is a machine in thermodynamic equilibrium, with no irreversible steps. But if a machine is reversible, it can just as easily move one way or the other. If it can move just as easily forward as it can move backward, it cannot do any useful work. For a machine to do useful work, we need irreversibility. This was the missing something we were looking for in Chapter 5.

In Chapter 3, we found that living beings are open, dissipative, near-equilibrium systems. Now this statement takes on a new meaning: The irreversible steps needed to put our cellular machinery to work are paid for by a continuous supply of free energy. We must receive free (low-entropy) energy from the outside (food or sunlight), and this free energy is degraded (dissipated) by our molecular machines as they use it to harness the molecular storm.

What Molecular Machines Eat

Molecular machines need a supply of free energy. In some sense, they eat free energy. But how do they like their free energy served? On a bun with some ketchup? Joking aside, in animals, the free energy that feeds the molecular machines of cells comes from food. A bewildering network of enzymes in the stomach, intestines, and cells breaks down food as part of metabolism. The final product of this complicated process is a molecule called adenosine triphosphate, or ATP, the energy-storage molecule that brought myosin to life in the motility assay mentioned in Chapter 4. Three phosphates bind to adenosine to form ATP. With all three phosphates attached, ATP is a bundle of concentrated energy. Snapping off one or two of the phosphate groups releases a great deal of energy—only the molecule's activation barrier keeps the phosphates from detaching right away. But once ATP binds to a molecular machine, the phosphate groups snap off readily, ATP turns into ADP (adenosine *di*-phosphate), and the machine is provided with a large amount of energy.

What form does this energy take? It is vibrational energy; the release of the phosphate makes the enzyme shake and rattle. In some sense, we can think of it as local heating (higher temperature means more violent motion). The energy released by the loss of one phosphate is equivalent to heating the enzyme up to 7,000 degrees Fahrenheit. This additional shaking allows the molecular machine to overcome activation barriers that are otherwise unattainable.

Once the ATP is broken down to ADP, the ADP goes back to the cell's recharging station, the mitochondrion. The mitochondrion is a cellular factory where sugar provides the energy to reattach phosphates to ADP, reconstituting ATP.

Tighty and Loosey

Righty tighty, lefty loosey

—GOOD ADVICE WHEN PUTTING TOGETHER IKEA FURNITURE

Now that we know what is on a molecular machine's menu, let us find out how real molecular machines implement the nano-Sisyphus method outlined above. As described earlier, nature has taken two major approaches to machine design: a robust ("super-Hummer") strategy and a floppy ("If you can't beat 'em, join 'em") strategy. Scientists refer to these as tight and loose coupling, respectively. The boundaries between the two are not well defined and are vigorously debated in the scientific community. Each time a new molecular machine is identified, the tight-versus-loose debate flares up again, with both factions claiming the new machine for their camp. Before I try to weigh in on this debate, let me further explain the difference between the two mechanisms.

When biophysicists talk of tight coupling, they are referring to two (related) ideas. First, they think of a close coupling between the supplied ATP and the work steps taken. Second, tight coupling refers to the machine's being bound to a molecular track at all times. A tightly coupled motor uses one ATP molecule to make exactly one step while at least partially bound to a track. A loosely coupled machine, by contrast, may let go of the track from time to time. Detaching from the track allows a loosely coupled machine to make more than one step per ATP molecule, but it also allows the molecular storm to push the machine in an undesired direction.

As discussed earlier, in the super-Hummer model (i.e., a tightly coupled machine), there must be some minimal degree of letting go—otherwise the machine would not move at all. A good analogy is a human walking. In order to move forward, we have to break contact with the ground from time to time (even when we shuffle, we shift from firm contact to sliding). We can do this because we always keep one foot firmly planted on the ground, while moving the other. This is, in essence, how a number of molecular machines, or motors, work. For example, a molecular motor called kinesin—a fifty-nanometer-long assembly of protein molecules—walks on two feet (or *heads*, as biologists confusingly call them) on a molecular track called a microtubule, always keeping one foot planted to the track. These motors are used to move cargo throughout cells—they are nanosize Sherpas, carrying heavy molecular loads along a one-way track to distant regions of the cell (Figure 6.8).

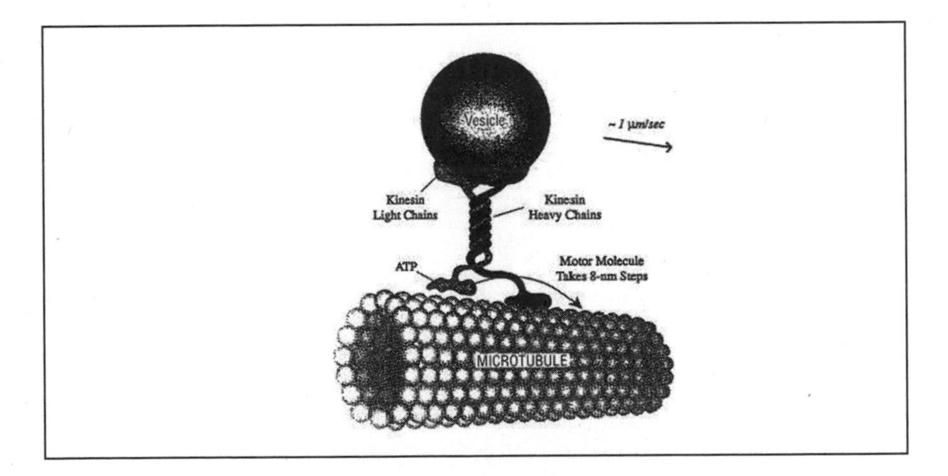

FIGURE 6.8. A kinesin molecular motor carrying a vesicle filled with nutrients walks along a microtubule. The whole molecular motor is only about fifty nanometers in height. © 1999 Robert A. Freitas Jr., www.nanomedicine.com. Used with permission.

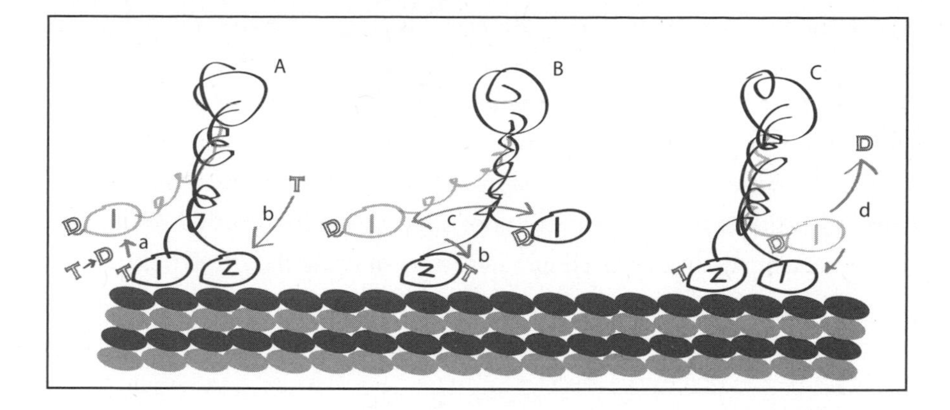

FIGURE 6.9. Motion of a kinesin molecule (highly simplified). A: Foot 2 binds an ATP molecule, while foot 1 degrades ATP (T) to ADP (D) and detaches from the track. B: Binding ATP (T) causes foot 2 to clamp on tightly to the microtubule track and allosterically bend its leg forward. Meanwhile, foot 1 has released from the track (arrow a in step A) and dangles back and forth (arrow c). C: But because foot 2 and its leg have tilted the molecule forward, forward motion is favored for foot 1, and eventually it will weakly bind to the track in front of foot 2, releasing ADP, as seen in arrow d. The molecule has taken a step. Finally (not shown), foot 1 binds an ATP, clamps on tightly to the track, and tilts its leg forward. The cycle repeats with the roles of feet 1 and 2 reversed: Foot 2 degrades its ATP to ADP and detaches from the track, and so on.

To see how kinesin motors work, let's take it step-by-step (literally). Initially, both feet are attached to the microtubule, and one of them has bound to an ATP molecule. The ATP molecule releases its energy (turning into ADP), and this energy is used to detach the foot from the microtubule. Once this foot is detached, the molecular storm initially pushes the foot forward and backward, but it cannot go very far as the other foot is still holding on to the track. Now, the foot still planted on the track takes on an ATP molecule and by an allosteric shape change bends the whole kinesin molecule forward, forcing the dangling foot toward the forward direction. Now, the dangling foot latches on in front of the attached foot, and releases ADP. The attached foot degrades its ATP in turn, detaches from the track, and the cycle repeats (Figure 6.9).

We find here a mechanism that is similar to our nano-Sisyphus model: The allosteric interaction plays the role of Sisyphus, not allowing the free foot to swing backward as it is randomly pushed by the molecular storm.

Once the foot has made a step, the now lagging foot must be released in an irreversible reset step, requiring the breakdown of an ATP molecule to a lower-energy ADP molecule. As long as there is a supply of ATP molecules, the motor will keep walking.

The hallmark of a tightly coupled molecular motor is that it goes through well-defined cycles, using up a fixed number of ATP molecules during each step. Nevertheless, random motion is the drive behind the motor's locomotion, as it ultimately moves the legs of the motor forward—of course, rectified by the allosteric interaction of the motor's legs with ATP.

Loosely coupled motors, by contrast, rely more heavily on random motion and have no fixed-step cycle. And they are more difficult to understand.

What Are the Odds?

To understand loosely coupled motors, consider two ways that nanoscale objects can move while immersed in water. We have already encountered random motion, which is caused by collisions with fast-moving water molecules. Physicists call this random motion *diffusion* (Figure 6.10). When a molecule moves by pure diffusion, it executes a random walk: Each step it takes is in a random direction, independent of the step it took previously. As a result, the molecule wanders around aimlessly, like a drunk after a night of binge drinking (sometimes this motion is called the *drunkard's walk*).

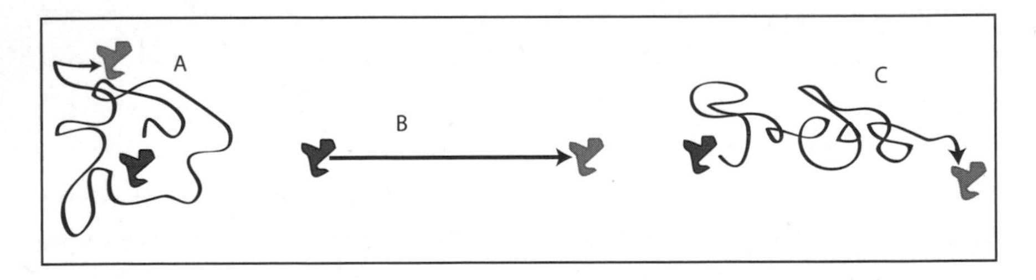

FIGURE 6.10. Possible molecular motions: (A) Pure diffusion: The molecule moves along a random path and makes no net headway. (B) Pure drift: The molecule moves along a straight, deterministic path in the direction of an applied force. (C) Drift and diffusion: The molecule moves, on average, in the direction of the force, but superimposed is random, diffusive motion.

The other type of motion occurs when a force is applied to the molecule. The molecule is still subject to random collisions from the molecular storm, but now, it is also dragged through the molecular tempest by the applied force. Physicists call this type of motion *drift*. What kind of forces can be applied to molecules? Ultimately, all forces in the molecular realm are of electrostatic origin. Even a chemical bond is ultimately due to electrostatic attraction (although some quantum mechanics is needed to fully explain the origin of a chemical bond).

An important fact to remember is that forces drive systems toward reduced energy. If a molecule binds to another molecule, the two molecules will not initially bind perfectly to each other—there will still be some distance between them. But as soon as the molecules feel one another, their combined energy will start to reduce, and they'll be compelled by a force of attraction. This force pulls the molecules in the direction of further energy reduction, that is, in the direction of completing the bond.

In general, molecules experience both types of motion, diffusion and drift. Drift is a deterministic motion, always in the direction of reduced energy, while diffusion adds a random component to the molecule's path (Figure 6.10). Because of this randomness, the best way to describe the motion of a molecule is to use statistical mechanics. The statement "in three seconds the molecule will move five nanometers to the left" becomes nonsensical, and we can merely state the *probability* that it may have moved to this position. Physicists can calculate the probability of a molecule's moving to a certain location by using the so-called Fokker-Planck equation (a challenging equation usually solved by computers). This equation calculates the probability of motion on the basis of the energy landscape of the molecule. The energy landscape gives a conceptual picture of the molecule's energy as a function of position and configuration. A steep slope in the landscape—that is, a strong dip in energy for small movements of the molecule—leads to drift. This is because steep changes in energy correspond to strong forces driving the system down the energy slope. On the other hand, on the flat parts of the landscape (where changes in the molecule's position or shape do not change energy by much), forces are negligible, and diffusion will dominate.

With this picture in mind, we can understand how the loosely coupled motor works. A loosely coupled motor is one that relies partly on random motion (diffusion), yet is still capable of performing directed motion. It can do this if it periodically attaches and detaches to a track. When it is attached to a track, it is subject to drift and feels a force directing it to the lowest-energy position on the track. When it is detached from the track, it diffuses freely. Loose coupling doesn't violate the second law, because detaching from the track requires free energy, which is degraded into heat. The detachment is the irreversible reset step.

How would such a machine move? We need to think of probabilities. Figure 6.11 shows a molecular motor on a track with an asymmetric energy landscape. The energy changes more steeply in one direction than the other, forming a saw-tooth pattern (similar to the teeth of the ratchet). On such an energy landscape, the combination of drift and diffusion will increase the probability that the machine will move one way rather than the other.

The asymmetry of the track's energy has two effects on the machine: First, the machine will spend most of its time near the energy minimum of the track. This minimum is at the lowest point of the track (Figure 6.11, step A). Now, when the machine detaches, it is just a short distance to the neighboring tooth on the left, but a larger distance to the tooth on the right. Second, once the motor reattaches to a random spot on the track, it will most likely end up on a gentle slope of the energy. This slope is also directed to the left, so the motor will experience a force pushing it to the left toward the minimum of the next tooth's energy. The overall result is that the asymmetry of the track's energy biases the motion of the motor to the left.

In Figure 6.11, the bottom three graphs represent this motion by plots of probability, each plot corresponding to a step in the cartoon above. Initially, if we know the motor is attached at a certain energy minimum, the probability that it is at this location is 100 percent (plot A). Then, the motor detaches and freely diffuses. This is represented by the flattened-out peak in the probability shown in plot B. When the motor reattaches, it quickly finds the nearest energy minimum. Because it was closer to the minimum to the left of its original position, the probability is somewhat higher that it ended up moving left rather than right. This represented in plot C by the slightly larger peaks in the probability to the left of the starting position. Note that the largest probability peak is at the starting position, indicating

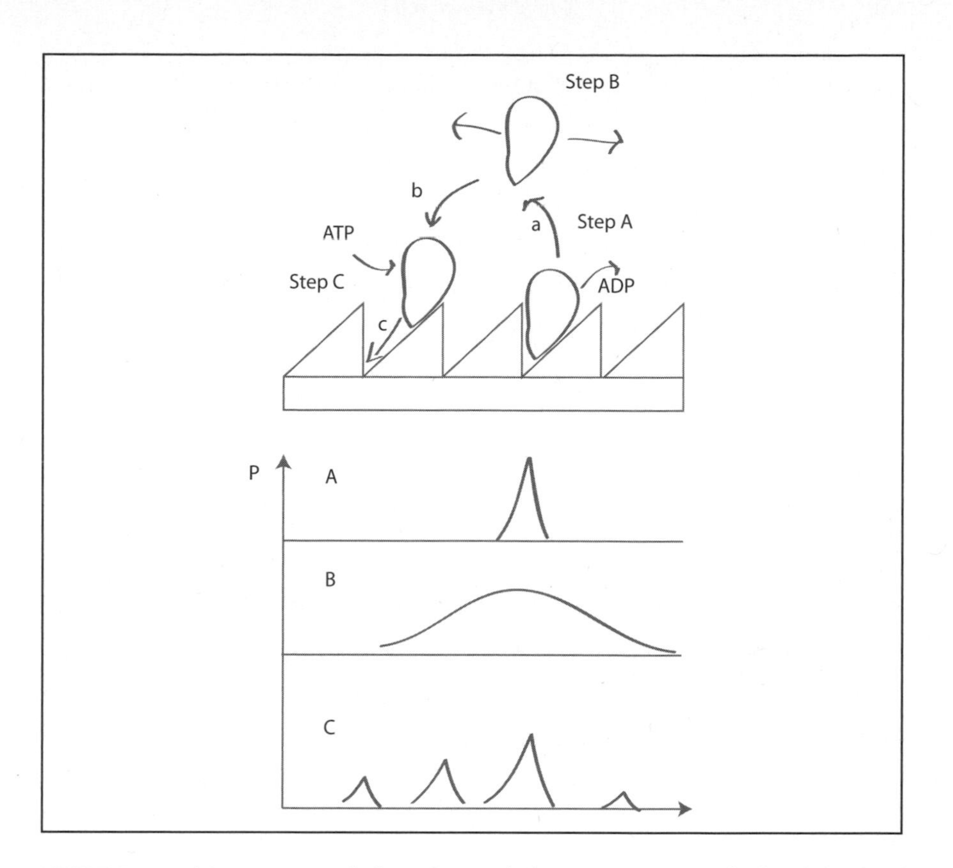

FIGURE 6.11. Top: Motion of a loosely coupled motor. In step A, the breakdown of an ATP molecule and release of ADP detaches the motor from the track (arrow a). In step B, the motor diffuses freely. In step C, the motor reattaches (arrow b), binding ATP in the process. As it binds to the molecular track, it quickly moves to the lowest energy state by drift (arrow c). Bottom: Probability plots corresponding to the steps shown above. (A) A single probability peak corresponds to the motor initially positioned at the energy minimum (prior to detaching in step A). (B) The motor diffuses freely, leading to a broadening of the probability peak. (C) The motor reattaches. Because of the asymmetry of the track's energy, the probability that it attaches to the left of the starting position is higher than the probability that it attaches to the right. The motor can also reattach at the same position where it started. However, leftward motion is more probable than rightward motion, and, on average, the motor will move to the left.

that often the motor does not move at all, but reattaches at its original location. There is also a small peak to the *right* of the starting position, indicating the small, but nonnegligible probability that the motor moves backward. On average, however, after many steps, the motor will move to the left. That is, it will perform directed motion.

If this model sounds eerily similar to our ratchet in Chapter 5, consider that physicists refer to this type of loosely coupled machine as a *Brownian ratchet*. The Brownian ratchet, unlike Feynman's passive ratchet, uses an irreversible step: the detachment from the track, fueled by the breakdown of an ATP molecule. We could, in fact, make Feynman's ratchet work, if from time to time, we injected energy to loosen and then retighten the pawl's spring. On loosening the spring, the wheel would rotate freely, with a slightly higher probability of rotating one way rather than the other. Tightening the pawl's spring would push the wheel further in the direction we want. On average, the wheel would move forward and do work. In fact, it can be shown that *any* molecular machine that operates on an asymmetric energy landscape and incorporates an irreversible, energy-degrading step can extract useful work from the molecular storm.

Because the motion of a loosely coupled machine is dominated by diffusion, the machine can take several steps at once, but it can also move backward. Therefore, loosely coupled machines do not move as efficiently as tightly coupled machines. Moreover, they do not move along a track for long distances before completely detaching. A motor that can move long distances along a track, continuously stepping forward, is called a processive motor. Tightly coupled motors tend to be processive; loosely coupled motors are not.

The Controversy

As mentioned earlier, there is an ongoing battle between scientists about whether molecular motors are tightly or loosely coupled. This fight has a deeper origin: Some scientists like the idea that the randomness of thermal motion plays a major role in molecular motors, while others prefer a more deterministic view. Loosely coupled motors have a clear dependence on random motion: The actual stepping is primarily done by random diffusion, with the asymmetric energy of the track providing the needed forward bias. Tightly coupled motors do not obviously incorporate random diffusion, so the tight-coupling mechanism is usually favored by many in the antirandomness camp.

A particularly intense debate is over the molecular motor myosin II. There are approximately eighteen different myosins in nature, and myosin II molecules make animal muscles work. When we decide to lift an arm, we send signals to countless myosin motors in our muscles and tell them to pull on actin fibers. Myosins grab the fibers and pull them taught, as in a game of tug-of-war. As the myosin motors pull in the fibers, our muscles contract and our arm lifts up. Next time you lift an arm (or move any part of your body), remember that this is accomplished by an army of myosin nanobots.

Measuring how these motors work at the single-molecule level is challenging, and unsurprisingly, different results have emerged from different laboratories. This has fueled the controversy. Until recently, most researchers believed that myosin II was a tightly coupled motor, making measured steps of five to ten nanometers per ATP molecule consumed. However, one of the pioneers of single molecular motors, Toshio Yanagida of Osaka University, produced repeated measurements that contradicted this claim. Yanagida claims that the distance myosin II moves fluctuates and can reach up to thirty nanometers per ATP molecule. This could only be possible if myosin were a loosely coupled motor.

Whichever way this controversy is resolved, the underlying debate over the relative role of the random thermal noise of the molecular storm is somewhat artificial. Even the most tightly coupled motor has to maneuver in the violent environment of the molecular storm. As we saw, a super-Hummer machine cannot work without occasionally loosening its grip. Once a molecular machine lets go, it is subjected to the random forces of thermal motion. A machine that can harness this random motion will be more efficient than a machine that cannot. As demonstrated, kinesin, a tightly coupled motor, uses the molecular storm to push its feet forward. The allosteric tilting of the molecule helps bias the movement in the forward direction, but where does the tilt come from? Any change in shape of a molecule is ultimately the result of the molecular storm's pushing the molecule in the direction of reduced energy, that is, into a valley of its energy landscape. A molecular motor will simply not work if

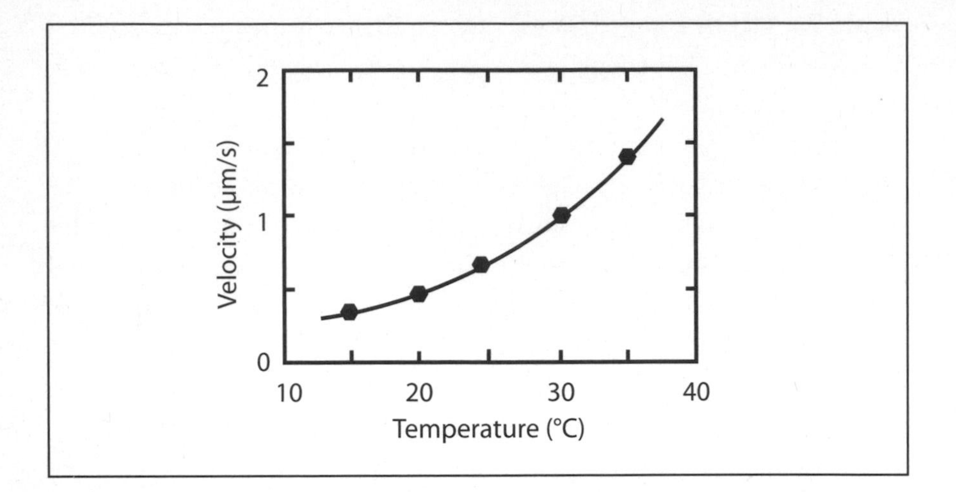

FIGURE 6.12. Speed of a molecular motor as a function of temperature (reprinted from Kenji Kawaguchi and Shin'ichi Ishiwata, "Temperature Dependence of Force, Velocity, and Processivity of Single Kinesin Molecules," *Biochemical and Biophysical Research Communications* 272, no. 3 (2000): 895–899, with permission from Elsevier). The speed increases exponentially as temperature is increased—a behavior typical for chemical transformations that need thermal motion to overcome transition barriers.

the temperature is too low to provide sufficient random thermal motion (Figure 6.12). Even the most tightly controlled motor needs the chaos of the thermal dance to traverse transition states and find its way on an ever-changing energy landscape. Without the chaos of the molecular storm, the molecular motors in our cells would not move and we would be dead.

7
Twist and Route

> Nature shows that molecules can serve as machines because living things work by means of such machinery. Enzymes are molecular machines that make, break, and rearrange the bonds holding other molecules together. Muscles are driven by molecular machines that haul fibers past one another. DNA serves as a data-storage system, transmitting digital instructions to molecular machines, the ribosomes, that manufacture protein molecules.
>
> —K. ERIC DREXLER, "MACHINES OF INNER SPACE," IN *NANOTECHNOLOGY: RESEARCH AND PERSPECTIVES*

THE CELL IS LIKE A CITY. THERE IS A LIBRARY (THE NUCLEUS, which contains the genetic material), power plants (mitochondria), highways (microtubules and actin filaments), trucks (kinesin and dynein), garbage disposals (lysosomes), city walls (membranes), post offices (Golgi apparatus), and many other structures fulfilling vital functions (Figure 7.1). All of these functions are performed by molecular machines. Some machines twist DNA; some route cargo along molecular highways or through the cell membrane. We now know that molecular machines are like tiny Maxwell's demons—on a steady diet of ATP. We now have some inkling of how they work. But there are numerous such machines in living

Quotation courtesy of K. Eric Drexler.

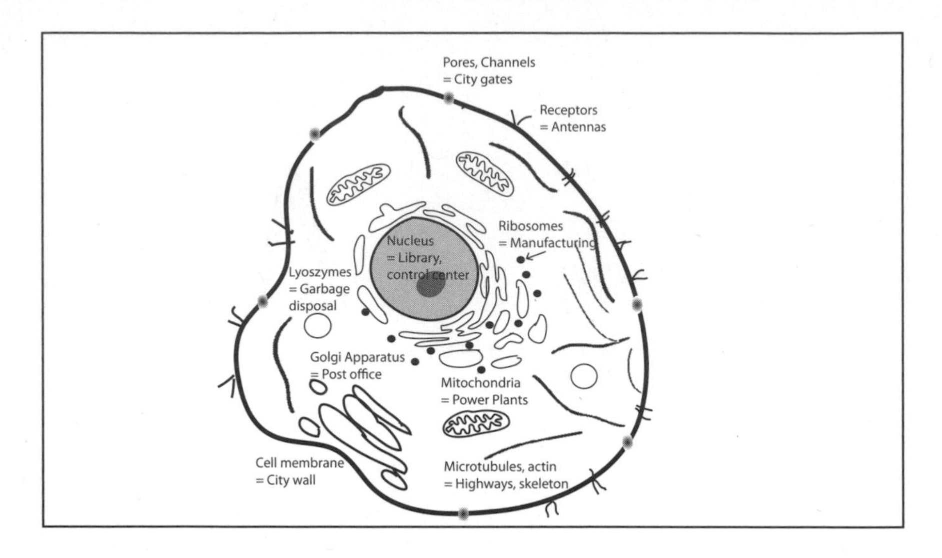

FIGURE 7.1. Schematic of an animal cell.

cells, doing many different things, moving in different ways, working together in ways we still do not fully understand. In this chapter, we will look at some of these machines, find out what they do, why we need them, how they work, how scientists have unraveled their mysteries—and the questions that remain. We will find that life does not yield its secrets so easily.

Walking the Walk: Kinesin, Myosin, and Dynein

In every city, there are menial jobs: lifting, carrying, rearranging, rebuilding. In the cell city, it's no different. Most of these jobs are done by three types of molecular machines: kinesin, myosin, and dynein (Figure 7.2), which we will call molecular motors or motor proteins, from now on. There is not one type of kinesin, myosin, or dynein doing one type of job. Instead, like a fleet of customizable trucks, there are superfamilies of molecular motors, with eighteen known classes of myosins, ten classes of kinesins, and two classes of dyneins. Each class, in turn, has many members, resulting in a giant number of variations on a common theme. Humans have genes that encode something like 150 different kinesins, myosins, and dyneins, including 40 to 50 myosins alone. On the other hand, the humble

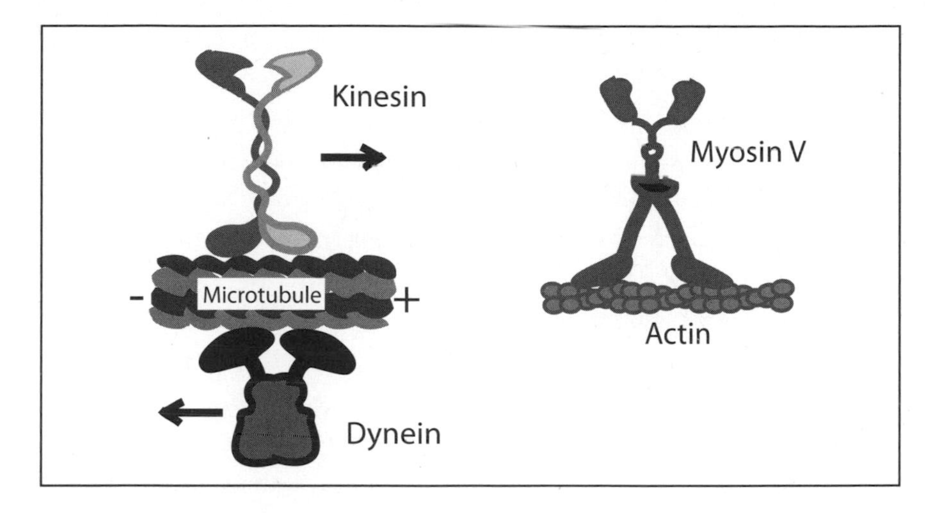

FIGURE 7.2. Molecular transport motors in our cells. These walking molecules are about thirty to fifty nanometers in height.

yeast fermenting your beer gets away with 5 myosins, which is still a lot for a single-celled organism.

The central part of each molecular motor is an ATPase. The suffix *-ase* signifies an enzyme, that is, a protein that facilitates a chemical reaction. In general, if the name of the enzyme is something like *X-ase* (pronounced "eks-ase"), the enzyme breaks down *X*. If the enzyme instead assists in the assembly of *X*, it is called an *X-synthase*. ATP synthase assembles ATP, while ATPase breaks down ATP. As discussed Chapter 6, the breakdown of ATP liberates a large amount of energy, which contributes most of the energy that drives a molecular motor. The ATPase site of the molecular motor is therefore the motor's combustion chamber.

The combustion chamber, a molecular pocket, snugly fits one ATP molecule. This ATPase pocket facilitates *hydrolysis*, the breaking off of a phosphate group from ATP, which leaves behind ADP. Hydrolysis (the phosphate is removed when ATP interacts with a water molecule) liberates 5.8×10^{-20} joules in energy, or 0.36 eV. This doesn't sound like a lot, but it is actually a fair amount of energy—about fourteen times the average energy per molecule contained in the molecular storm. Such a high energy corresponds to locally heating a molecule to 3,900 degrees Celsius, or 7,000 degrees Fahrenheit. The hydrolysis of ATP also produces a small change in the shape of the ATPase pocket, creating a displacement of about 0.5 nm. Knowing the energy and the displacement, we can calculate the maximum force that can be generated by the hydrolysis of ATP in the ATPase pocket: Assuming 100 percent efficiency, or assuming that all the released energy is available to do work, the force would equal the released energy divided by the displacement, or 5.8×10^{-20} joules, divided by 0.5×10^{-9} m (1 nm = 10^{-9} m), which yields a force of 116 piconewtons (pN) (*pico* denotes one-trillionth, or 10^{-12}). How big of a force is this? If we compare a force of 116 pN to the weight of the smallest visible dust particle, the weight of the dust particle is still larger by about a factor of 5. The energy released from one ATP molecule could lift the dust particle by a mere 0.1 nm, or the size of a hydrogen atom. This sounds incredibly tiny, but in the molecular world of our cells, 116 pN is a huge force. For example, it would be sufficient to rupture the membrane of the cell.

While the force is large, the movement is tiny. It's difficult to do a whole lot of useful work when moving in 0.5-nm steps. For this reason, the ATPase pocket in molecular motors is coupled to a mechanical amplifier—a part of the molecule that, like a lever, amplifies the motion of the ATPase pocket into a larger displacement. We encountered such mechanical amplification before when we discussed allostery. Energy conservation demands that the force is reduced proportionally to the increase in displacement. If I want to generate ten times the motion, specifically, 5 nm, the force I could generate would be ten times smaller, or only 11.6 pN. In reality, the efficiency of these motors is only about 60 percent (which is still very good), and amplified motions range from 5 to 36 nm, while forces range from 1 to 10 pN. This really does seem small, but even 1 pN is enough to move molecular cargo around, especially if a number of motors work together.

TWEEZERS OF LIGHT

Measuring the displacements and forces generated by molecular motors sounds like an incredibly tricky proposition. After all, these motors are only a few tens of nanometers large, and we have seen how incredibly tiny a nanometer is. You could hardly put one of these tiny molecules on a leash and measure how much it tugs. Or could you?

In 1993, Steven Block and his research group at Harvard attached kinesin molecular motors to small silica beads, using small molecular leashes. The beads were then captured in a trap made of light—a laser trap. Light is made of particles, or photons, and these photons have momentum and can impart an impulse on a particle. Many photons together can create pressure. In a laser trap, high-intensity laser light is focused such that the pressure of the photons creates a minute inward force, a force that will keep a small particle from escaping the center of the laser focus. The force is only about 1 to 10 pN, but the weight of a 600-nm diameter bead, as used in Block's study, is only 0.002 pN, so the light pressure is more than sufficient to suspend and hold the bead (Figure 7.3). Laser traps are amazing devices: It is relatively easy to move light with lenses and mirrors and thus the focus and the trapped bead can be moved together in three dimensions. The laser trap thus acts like highly sensitive tweezers—tweezers that can pick up nanometer-size beads and single molecules attached to the bead.

In Block's experiment, he and his students picked up a bead out of a suspension of numerous beads. Each bead was decorated with kinesin molecules. Once a bead was captured, it was lowered onto a microtubule, the protein track on which kinesin molecular motors move, which was fixed to a glass slide. Block's group combined their laser trap with interferometry, using two separate light beams to illuminate the object. From the interference of the light scattered by the bead, they could determine the motion of the bead with high accuracy.

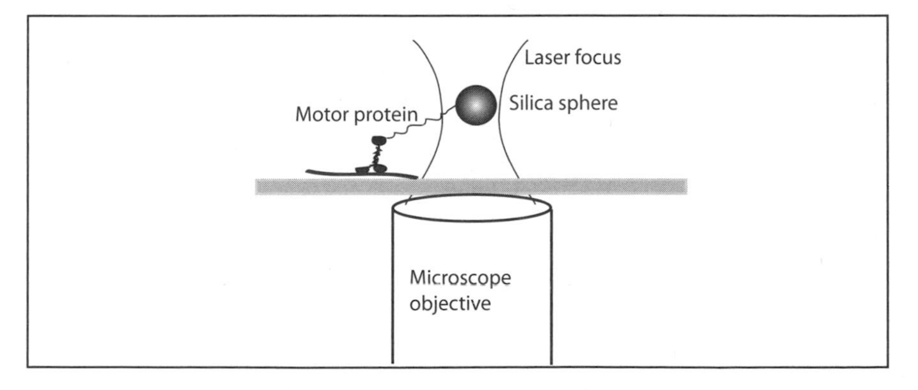

FIGURE 7.3. A laser tweezer measurement of the motion and forces of a transport motor protein.

Motion came from two sources: thermal noise (the molecular storm), and the deliberate motion of the motor proteins. As the kinesin molecule moved along its track, it pulled the bead along with it. As the bead moved to the edge of the light trap, the force increased from 0 to 1.5 pN. Tracking the motion and simultaneously measuring the force, Block and his colleagues were able to measure the speed and step size of a single kinesin motor as a function of the load it had to pull. Once the force became too much, the motor detached from the track, the bead was pulled back into the center of laser focus, the kinesin motor attached again, and the measurement was repeated.

At low ATP concentrations, Block found that the bead was pulled by the motor with a speed of 50 nm per second. Applying statistics to the noisy data, they found that kinesin took 6- to 8-nm-size steps, and therefore the kinesin seemed to step about six to eight times every second. They also found that the kinesin was an almost perfect rectifier of thermal motion, almost never stepping backward.

As they raised the ATP concentration, the motor became faster, finally reaching a maximum speed of 500 nm per second. At this speed, the little molecule would traverse an entire cell in about a minute. Now, Block and his students gave the little motor a challenge: Increasing the laser intensity, they created a higher trapping force and made the little motor pull against it. They found that the kinesin motor could pull against forces as high as 5 pN. By comparison, the force needed to pull a 600-nm bead at 500 nm per second through water is about 0.03 pN. Thus, this little 50-nm-long motor can easily pull an object many times its size at a pretty fast clip through the cell. It is a molecular worker ant.

Since Block's pioneering work, the use of laser tweezers to study molecular machines has advanced considerably. Noise is much lower, feedback loops are used to control force or position, and multiple traps can manipulate single molecules in myriad ways. Sixteen years later, Block and his coworkers have been able to follow the movement of a single leg of kinesin by attaching the bead to one of the heads of the motor protein via a DNA tether. Using the DNA tether, they gently pulled on the leg to see if it was bound to the microtubule or if it was dangling freely. In addition to the amazing skill involved in chemically attaching a DNA molecule to a specific 10-nm part of a 50-nm molecule and then attaching the DNA to a bead trapped in a beam of laser light, the data obtained greatly surpassed Block's 1993 data. Back then, Block and his students had to use statistical analysis to extract meaningful information from the noisy traces of the bead's movement. In 2009, every step of the molecules head was clearly seen with nanometer resolution.

FLUORESCENCE MAGIC

The amazing feats of molecular motors were being discovered long before laser tweezers were developed. The first true single-molecule techniques were developed in the 1980s. These methods were based on attaching fluorescent molecules to the motor protein. Fluorescence is the emission of characteristic colors of light in response to a material's being excited by light of higher energy. When high-energy photons hit a fluorescent molecule, they cause the electrons in the molecule to jump to a higher orbit (or energy state). The molecule is now in an excited state, but after some time, the electrons will relax back to a lower energy state. The difference in energy between the excited and the lower energy state is emitted as light of a characteristic energy and, therefore, a characteristic color. A well-known example of fluorescence is the eerie glow of your T-shirt when you enter a dance club with black lights. The ultraviolet light produced by black lights is highly energetic and can excite the molecules in your clothes. As the electrons fall back down, they produce a bluish visible light—this is fluorescence.

Using sensitive sensors, researchers could detect the light from a single fluorescence molecule. In this way, they could follow the motions of single molecular motors in an optical microscope, even if the motors themselves were too small to see. All the scientists had to do was follow the light emitted by the fluorescent marker attached to the motor.

This ability to detect amounts of light down to single photons has produced a cottage industry of fluorescence-based single-molecule techniques. As is customary in science, they all have fancy acronyms, from TIRF to FRET, FCS, and SHRIMP. A pioneer in using fluorescence techniques to study molecular motors is Jim Spudich at Stanford University. In 1983, he developed what is now known as the bead assay. Spudich and his students attached myosin molecules to fluorescently labeled beads (much as Block attached kinesin to beads ten years later for his laser trap). He then floated the beads onto a bed of actin filaments. As soon as the myosin molecules contacted the actin and were properly fed with ATP, the fluorescent beads started to move around like microscopic fireflies. In 1986, Spudich did the opposite. This time he attached myosin to a substrate and then floated fluorescently labeled actin filaments on top of the motors. Again, upon proper feeding, the myosin molecule started to push the actin filaments around, creating a wiggling spectacle of gliding filaments. This gliding-filament assay is what I saw twenty-five years later at the biophysical meeting, and it was still as fascinating as when it was first created. Watching the filaments move around like glowing worms reminded me that the motion inherent in life, so mysterious to the Greeks and early-twentieth-century researchers alike, was now without a doubt explained by the amazing motions of mere molecules.

HOW MOTORS STEP: KINESIN AND MYOSIN V

Kinesin-1 and myosin V (for some reason, the kinesin researchers like to use Arabic numerals, and the myosin researchers Roman) are the quintessential walking motors. Both these proteins are processive, so they can walk long distances without falling off their tracks. Kinesin walks on microtubules, while myosin strolls along actin filaments. Scores of detailed studies involving fluorescence, AFM, laser tweezers, X-rays, electron microscopy, and the use of mutated motor proteins have brought us very close to deciphering a detailed picture of how these amazing walking molecules move. Structural studies have shown that the ATP binding site, or *motor domain*—the part of the motor that binds its track and binds ATP—is very similar in both proteins. This indicates that the two come from a common ancestor. However, somewhat surprisingly, the way the two molecules move is completely different. While myosin V strides, kinesin-1 waddles. Moreover, the two motor proteins use their ATP for completely different purposes, as we will see.

The details of how the motors walk are still controversial. Let's look more closely at kinesin, or, rather, a specific kind of kinesin, called kinesin-1. We briefly observed kinesin in Chapter 6, to gain some understanding of the complex issues. Kinesin-1 consists of two motor domains (or *heads*),

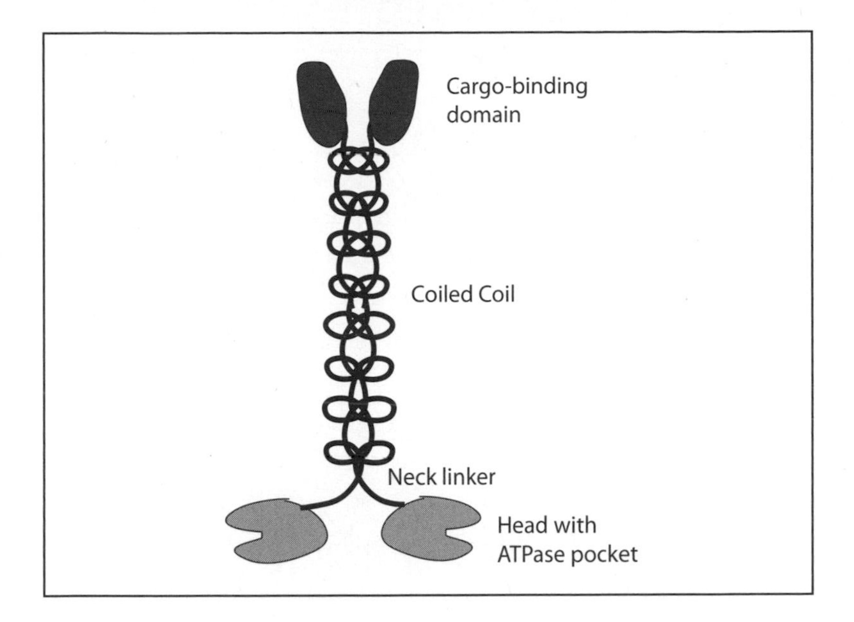

FIGURE 7.4. Structure of a kinesin-1 molecule. At the bottom are the two heads on which the molecule waddles along a microtubule filament. Each head has a pocket to bind and hydrolyze ATP. The two heads are held together by a short neck linker. On top of this is a coiled-coil stalk, which ends with the cargo-binding domain.

which attach to, and detach from, the microtubule. The heads are linked through a short neck linker, which does not allow a huge amount of flexibility. Thus kinesin is condemned to waddling—it's like how you might walk if your pants had slid down to your ankles. Beyond the neck linker is the *coiled coil* (that's not a typo—it's a coil of two coils), which is the longest part of the molecule. The coiled coil is crowned by a cargo-binding domain, whose function is self-explanatory (Figure 7.4).

To follow the motion of a kinesin molecule, we have to start at some point along its cycle. Let's start just as the head in front is empty, binding neither ATP nor ADP but bound to the microtubule. At the same moment, the trailing head has an ADP bound in its ATPase pocket. Block's 2009 experiment suggests (for now) that the ADP-bound head's affinity to the microtubule is very low. In other words, the trailing head is freely dangling, while the leading, empty head is firmly attached to the microtubule. Block calls this configuration the waiting state (Figure 7.5).

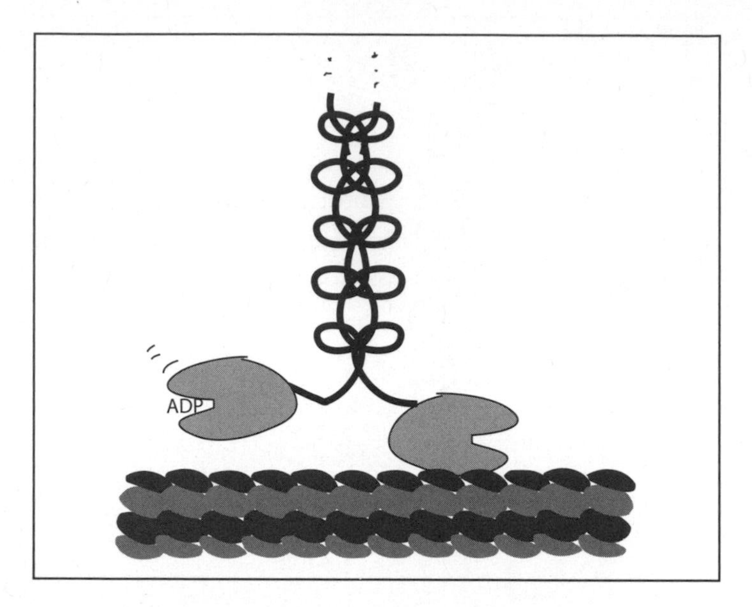

FIGURE 7.5. The waiting state of a kinesin motor protein on top of a microtubule. The cargo domain has been omitted for simplicity.

Now, an ATP molecule floats by. It can't bind to the trailing, dangling head (its binding pocket is already occupied by ADP), but the empty head in front is happy to take in the lonely ATP. Binding ATP releases energy, and in this case, the energy is used to dock the neck linker with the leading head. This tilts the molecule forward and produces some strain between the two heads (which, after all, are linked by a rather short neck). By a combination of the strain in the linker and, possibly, the tilted energy landscape provided by the microtubule, the dangling head performs what is called *biased diffusion*. That is, the head is pushed about by the molecular storm, and since the energy landscape is tilted in the forward direction, the head moves forward—most of the time (but not always—sometimes it will step backward!).

This diffusive search for a low-energy landing place is still a bit of a mystery. Some research groups have determined that the docking of the neck linker is associated with an energy that is comparable with the energy contained in the molecular storm. This would not be enough to resist the onslaught of marauding water molecules. However, other researchers found that ATP binding may provide a much higher amount of energy, around thirteen times the thermal energy of the molecular storm. Comfortable sites on which the head could land are separated by 8 nm. The site directly ahead is already occupied by the leading head, so the dangling head has to diffuse 16 nm to find a new, comfortable resting place, an appreciable distance for a molecular motor. The dangling head must step around the planted head, walking like a cowboy wearing chaps, to find its resting place. It is difficult to see how a small forward tilt of maybe 1–2 nm due to the docking could propel the head a full 16 nm. Yanagida's group, of myosin II controversy fame, measured how kinesin steps at different temperatures and found an entropy contribution worth four times the energy of the molecular storm. Thus, the missing component may be entropy; in other words, the forward motion is entropically favored over backward motion.

Once the searching head finds it new resting place, 16 nm from where it started, it releases its ADP and tightly binds to the microtubule. The empty binding pocket of this head is now ready to bind an ATP. The molecule has taken one step, the leading head is now trailing, and the formerly dangling head is now leading. But we are not yet back to square one: The now trailing head is still bound to ATP. Through an allosteric interaction, the ATP in the trailing head keeps the new leading head from binding an ATP. This is a good thing. If the leading head were to bind ATP while the rear head was still attached to the track, it would dock the neck linker and introduce intolerable strain in the molecule. Instead, the trailing head first hydrolyzes its ATP, splitting off a phosphate ion and turning ATP into ADP. Once this happens, the trailing head detaches, the leading head is free to bind ATP, and the cycle repeats (Figure 7.6). We can see from this sequence how stepping is choreographed: The motion controls the chemical transitions, and the chemical transitions control the motion through allostery. This chemomechanical coupling ensures that the motor never falls out of step. It is this tight control of the motion that makes kinesin a highly processive motor.

Let us relate this detailed scheme to the more general ideas introduced in the previous chapters. There we saw that we need a tilted energy landscape and a supply of free energy to reset the system. In kinesin, the tilted energy landscape comes from the strain induced between the two heads by the short neck linker, the binding energy of the microtubule, and, possibly,

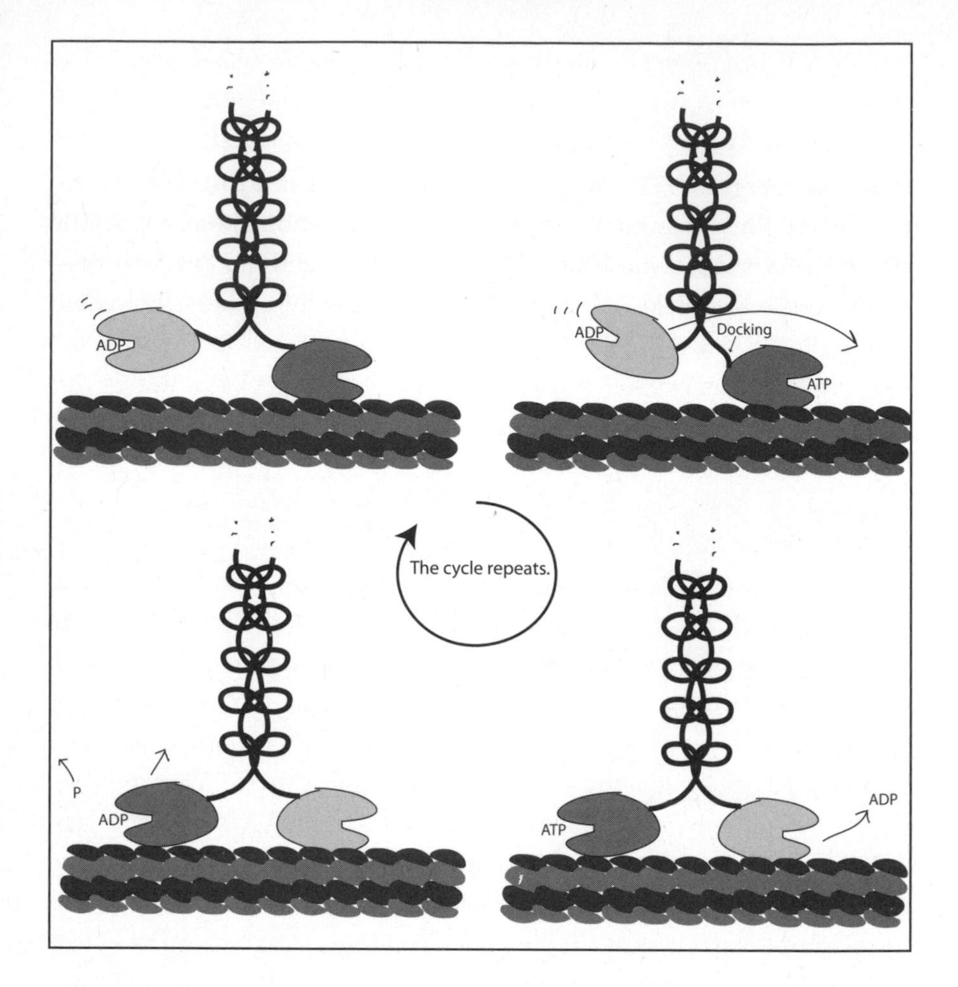

FIGURE 7.6. Left to right: The step-by-step motion of kinesin-1. At the end of the sequence, the cycle repeats, with the two heads having traded places and the molecular motor having advanced one step.

an entropy contribution. Free energy is supplied when ATP binds, when the head binds to the microtubule, and when ATP is hydrolyzed. Some of this energy is used to dock the neck linker and produce strain, which in turn imparts kinetic energy to the dangling head. Some energy must also be used to detach from the microtubule and to release spent ADP. Eventually all energy will lead to a vibration of the kinesin and microtubule and end up as waste heat. The result is that free energy, in the form of ATP, is used to do some work, and the rest of the energy is dissipated as heat. No violation of the second law here.

What is striking, however, is that the main source of energy, the hydrolysis of ATP, does not seem to be directly used for locomotion. Rather, as in the case of Maxwell's demon, or the nano-Sisyphus, it provides a reset step. The energy in ATP is used to detach the trailing head from the microtubule and thus, in some sense, erase the information about where the trailing head was previously attached. Kinesin is a tightly coupled motor—it takes one ATP to take one step. It holds on tight when it must, but when it steps, it acts like a Brownian ratchet, diffusing on an energy landscape that is sculpted not only by the environment, but also by mechanical strains generated inside the molecule itself. What an ingenious little machine!

THE FAMILY OF KINESINS

We have so far only talked about one type of kinesin: kinesin-1. As mentioned, there are ten classes of kinesins, and each class has many members. Humans have forty-five different kinesins encoded in our DNA. There are many types of kinesin because they all have different jobs and work a little bit differently. But they are all kinesins, because they share an identical motor domain and they are all molecular motors.

The location of the motor domain in relation to the rest of the protein varies greatly between the different kinesins. In general, proteins have two ends. Proteins are folded strings of amino acids, and molecular motors are no different. Each amino acid has a carboxyl group at one end and an amino group on the other. When the amino acids link together, the carboxyl end of one amino acid hooks up with the amino end of the other, forming what is called a peptide bond. But that leaves a carboxyl group at one end and an unbounded amino group at the other end. As more and more amino acids are added, there is always a carboxyl end (called the C terminus) and an amino end (called the N terminus—because amino groups contain nitrogen). Kinesins can be classed into three groups, depending on where the motor protein is located: near the C terminus, near the N terminus, or somewhere in between. Most kinesin molecules are N-terminal, but there are three C-terminal and three M (middle) kinesins in humans. Interestingly, N-terminal kinesins walk along microtubules toward what is called the plus end (toward the cell wall, away from the interior of the cell), while C-terminal kinesins walk toward the minus end (toward the cell center).

Kinesins transport vesicles (lipid-enclosed sacks), which can contain various molecules, or they create tension in microtubule networks, especially during cell division. Some kinesins are active only in the cells of our nervous system, and indeed, mutations to the amino acid sequence of these kinesins are linked to neurological diseases. Some kinesins move whole organelles (substantial subunits of the cell). For example, a kinesin called KIF1B moves mitochondria in nerve cells. As mentioned earlier, mitochondria are the cell's power plants, which recharge ADP to ATP. Other kinesins (e.g., kinesin-2), have nonidentical motor domains and play important roles in embryonic development. Chromokinesin is involved in moving DNA around during cell division. When cells divide, the DNA information has to be equitably distributed between the two daughter cells. This is achieved by separating the chromosomes (which are bundled up DNA) in a complicated process called mitosis. This separation is accomplished by armies of molecular machines, kinesins among them.

Two of the more bizarre kinesins are kinesin-3 and kinesin-5. Kinesin-3 appears to only have one head and a loop to hold on to a microtubule. Kinesin-3 has been suggested as an example of a Brownian ratchet–based motor, because a single head cannot really move the way kinesin-1 moves; how could strain be generated if no part of the molecule is attached to the microtubule? In this model, the loop may help to hang on to the microtubule while the head detaches for (directed) diffusion. However, recent research has suggested that kinesin-3 molecules pair up and work together, much like the two permanently connected heads in a kinesin-1 molecule.

Kinesin-5 represents the opposite extreme: It has four heads, two pairs back-to-back. Kinesin-5 is involved in mitosis. It attaches to two microtubules at the same time and pulls them together. During mitosis, the chromosomes are attached to a structure called the spindle, which is made of microtubules. Kinesin-5 is the motor that keeps the spindle taut.

MYOSIN V: RELATING MOTION AND CHEMISTRY

In 2009, about a year into a new undergraduate program—biomedical physics—at Wayne State University, our department hired a biophysicist. Through a stroke of luck, it turned out to be my colleague Takeshi

Sakamoto. Takeshi specializes in molecular motors, with several groundbreaking works to his credit. And his baby is myosin V.

Myosin V, like kinesin, is a walking transport motor. There are some differences, however: Myosin moves on actin filaments rather than microtubules, and it has long legs, rather than a short neck. Thus, it strides, rather than waddles. While kinesin and dynein are the long-distance trains of the cell, myosin V is more like a local transporter. It takes cargo from kinesin and moves it a short distance to the cell membrane, where the cargo is passed off to other molecular machines, which then move the cargo out of the cell.

At the 2011 Biophysical Society Meeting, I met Takeshi's Ph.D. advisor, who is from Kanazawa University. Toshio Ando blamed Takeshi for getting him hooked on myosin V. Ando was not too unhappy about this addiction. His fascination with molecular motors drove him to create one of the most sophisticated atomic force microscopes in the world. Ando's AFM can scan so fast that a single image is completed in twenty milliseconds. This may not mean much to the uninitiated, but AFMs usually take tens of seconds or even minutes to complete one image. As described earlier, AFMs create an image by moving a sharp tip over a surface, and if you move the tip too fast, you will tear your sample to shreds. To move it as fast as Ando does, you need superfast electronics and hardware. It took Ando and his students almost ten years to perfect their technique.

Ando showed his now-iconic movie of a walking myosin V molecule at the biophysical meeting. Using his high-speed AFM, Ando and his students filmed single myosin molecules as they walked along actin filaments. These are mesmerizing little movies: They are fuzzy and the images wobble, but you can easily see a nanometer-size machine, like a little two-legged creature, wait, then suddenly do a quick step, then wait again, step again, and so on. Is this molecule alive? No, not in the full sense of the word. But watching it stride by, you can see how many such machines, interacting in some regulated way, can make a living being. This surely is where life begins.

Takeshi left Ando's group before they made this movie, but he joined another well-known group studying myosin V: Jim Sellers's group at the National Institutes of Health in Bethesda, Maryland. Sellers had been studying myosins (especially myosin II, muscle myosin) since he completed his

Ph.D. in 1980. Takeshi worked in Sellers's group for eight years before joining Wayne State. His crowning achievement was the simultaneous imaging of the motion of a walking myosin V molecule and its uptake of ATP. This study was the holy grail: the linking of the biochemical cycle (ATP binding, hydrolysis, and ADP release) with the mechanical cycle (detaching, diffusing, stepping, reattaching). Linking the two in space and time on a single molecule goes a long way toward understanding how these molecules convert chemical energy into mechanical energy. This had never been done, not for kinesin or any of the other molecular motors.

To link the chemical and mechanical cycles, Takeshi used an ATP with an attached fluorescent label. This label shone up to twenty-five times brighter when the labeled ATP bound to myosin. Thus, Takeshi and coworkers could easily tell if the nucleotide (ATP or ADP) was bound to a myosin head. To follow the motion of the myosin, they used another label attached directly to the body of the myosin. Takeshi found that myosin V was a tightly coupled motor—one ATP, one step. Kinesin is also believed to be a tightly coupled motor, but this is still not completely proven. Thus, Takeshi's experiments were the first to clearly demonstrate, for any molecular machine, a one-to-one relationship between ATP hydrolysis and motion. This was an important finding, because after Yanagida's work, it seemed possible that myosin V's cousin, myosin II, was weakly coupled, moving variable distances for each ATP consumed.

When Takeshi set up his experiments, it was already known that myosin V—unlike kinesin—took large 36-nm long steps. But there were still unanswered questions, such as how motion related to the use of chemical energy. To establish this link, Takeshi and his coworkers first primed their myosin V with a fluorescent label and supplied it with fluorescently labeled ATP. After focusing their microscope, they observed bright dots of different colors (they could not see the molecules directly, but only observed the fluorescent labels) moving along actin filaments in a choreographed step dance. First the ATP dot moved 18 nm, then the myosin dot moved 36 nm, and then the ATP dot moved another 18 nm, and then the cycle repeated (Figure 7.7). What did this mean? Takeshi and his colleagues could not distinguish between ATP, ADP-P (an ATP molecule in the process of being hydrolyzed—basically ADP with a nearby phosphate group), or ADP. All forms of the nucleotide shone with the same color and

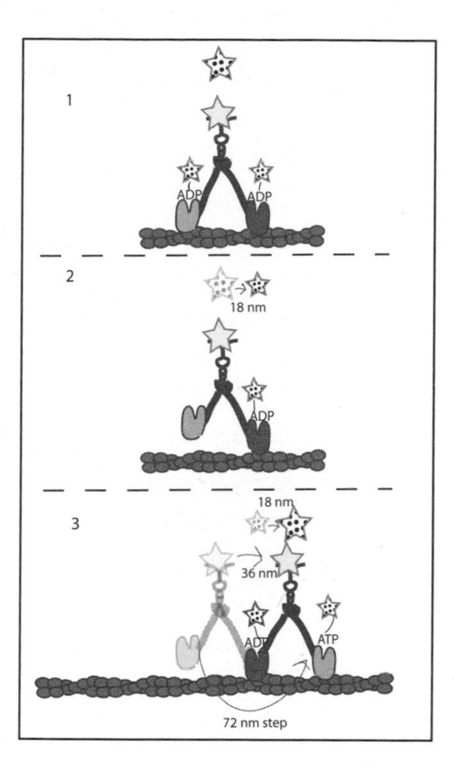

FIGURE 7.7. The motion of myosin V and the associated fluorescent labels in Takeshi Sakamoto's experiment. Fluorescent labels are shown as stars. Polka-dotted stars represent fluorescent labels attached to nucleotides or the visible fluorescent spot due to these labels (floating above each molecule). The size of the star indicates the brightness of the fluorescence. If both heads had nucleotides bound to them, one bright fluorescent spot was seen at the center of the myosin, while when only one nucleotide was bound to the myosin, a dimmer spot was seen at the location of the nucleotide bound head. The nondotted star represents the fluorescent label that is directly attached to the myosin molecule. Arrows indicate movement. See text for more explanation.

brightness. Another complication arose when both heads were bound to a nucleotide. In this case, the researchers did not see two separate fluorescent dots, but just one dot, albeit one twice as bright. This was because the resolution of their microscope was not sufficient to distinguish two fluorescent labels so close to each other.

With this information at hand, and with the knowledge gained from previous experiments, Takeshi and his coworkers surmised that the first step in the cycle was the dissociation of ADP from the trailing head. Initially, when ADP was bound to both heads, the brightest location of the fluorescence was between the two heads. But as one ADP dissociated and blinked out, only the front head remained illuminated by its label. Thus the spot they saw moved from the center of the molecule to the leading head, while dimming to half its brightness. Since the heads were separated by 36 nm, this meant the spot moved by half this distance, or 18 nm.

Next, the fluorescent label attached to the myosin moved by 36 nm. This meant that following the release of ADP from the trailing head, this head made a 72-nm step, hand-over-hand, to propel the whole molecule

36 nm (remember, the other head was still attached and did not move). Almost simultaneously with this 36-nm motion, the ATP label moved 18 nm. Why 18 nm and not 36? Because in the meantime, ATP had attached itself to the forward-moving head, thus moving the center of the nucleotide fluorescence back to the center of the molecule.

It had previously been discovered that a myosin head, when bound to ATP or ADP-P, binds very weakly to the actin filament, although it binds strongly to actin in the presence of ADP or if the head is empty. Thus, very rapidly, prior to rebinding to the filament, the forward-moving head hydrolyzed its ATP into ADP-P. Then it released P and, now left with only ADP in its pocket, latched on to the actin filament. The timing of these steps, which happen very rapidly, is still not completely understood. Is phosphate released prior to actin binding, or after? Is it released before or after a conformational change of the head, which creates the forward tilt? These questions remain unanswered.

An even greater mystery surrounds the use of ATP hydrolysis in myosin V. For one, myosin V is completely different from kinesin, where ATP hydrolysis was used to detach the head from the filament. In the case of myosin, ATP hydrolysis apparently leads to the *attachment* of the head to the filament. But attaching a head to the filament does not cost energy; it should supply energy. So why supply even more energy at this step by hydrolyzing ATP? This does not make sense to me. If not for binding, how does myosin use the energy supplied by ATP hydrolysis? At the writing of this book, this is still a mystery.

When asked about this, Takeshi said his hunch was that the energy of ATP hydrolysis is not used directly to produce motion. If this is correct, the energy to provide motion comes from the energy of binding to the track, the release of internal strain in the molecule, and the thermal energy of the molecular storm. But then myosin would violate the second law. Thus, ATP hydrolysis needs to play the role of the reset step. For example, hydrolysis might prime a power stroke, the deformation that tilts the molecule forward to induce the trailing head to move in the forward direction. As the moving head swings forward, its long neck is oriented vertically. When ATP hydrolyzes, apparently just before binding to actin, the neck linker swings backward, releasing strain in the molecule. Upon binding, the neck linker swings forward again (probably powered by the binding energy when the head binds to actin), providing the strain that tilts the molecule in the forward direction before it can take another step. So, again, the free energy supplied by ATP hydrolysis is used to reset the system. In the case of myosin, it may be used to temporarily release strain, to allow it to be applied again as soon as the molecule is in a position where this strain becomes useful to rectify its motion.

MYOSIN II: MUSCLE MOTION

Myosin V is no doubt fascinating due to its iconic, long-legged walking motion. But the original myosin, and the one most linked to the motions of animals that so fascinated biologists from Aristotle onward, is myosin II. The molecular machine that makes muscles work, myosin II was the first molecular motor discovered—by Andrew Huxley, in 1969. Strictly speaking, in evolutionary history, myosin II existed before muscles. "Nonmuscle" myosin II in cells pulls on actin filaments, which are part of the cell's skeleton, and thereby helps the cells change their shape. As is common in evolution, once multicellular organisms evolved, the actin-pulling machines were co-opted to move the whole organism—using muscles.

It is quite easy to see how nonmuscle myosin may have turned into muscle myosin: Muscles are made of bundled bundles of muscle fibers—each fiber a single cell. Each cell contains myofibrils, which are structures composed of actin filaments interlaced with fibrous bundles of myosin II (Figure 7.8). Muscle cells are cells with an exaggerated network of actin and myosin—otherwise, myosin does what it has always done, even before muscles existed: pull on actin.

The structure of muscle and the presence of a molecule called myosin was known long before 1969, when Huxley proposed the first model to explain how an army of molecular motors could produce macroscopic motion. At the time, it was already known that muscle contraction was due to the relative sliding of two different filaments against each other: actin filaments and myosin filaments. But what generated the force that made the two filaments move relative to each other? Using electron microscopy of muscle fibers, Huxley suggested that there were distinct crossbridges between the two filaments and that these bridges generated force and motion by tilting. The crossbridges turned out to be myosin heads,

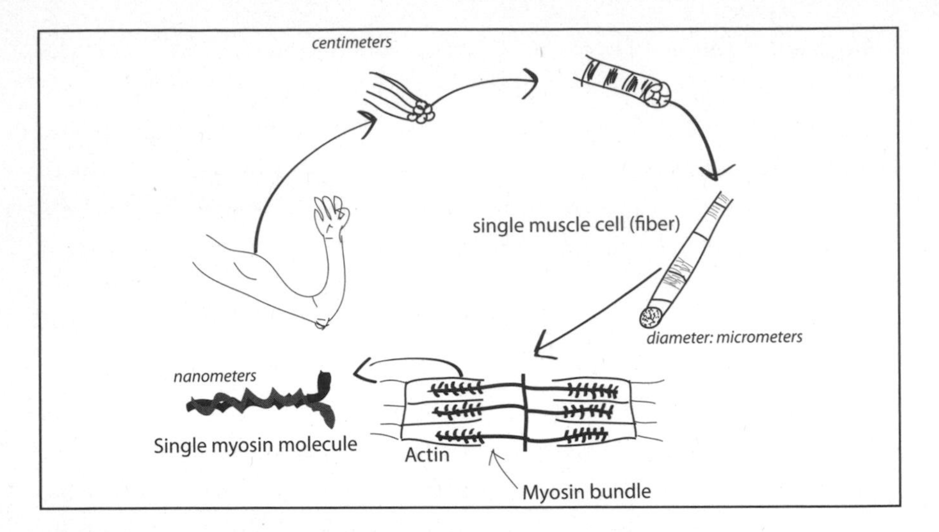

FIGURE 7.8. The structure of muscle.

and thus muscle motion was now attributed to the concerted efforts of legions of myosin molecules pulling on actin filaments.

It is worth considering the scales involved in the various components of muscular movement. Muscles are macroscopic objects with a typical dimension of centimeters (or inches). As we zoom in, we first find that muscles are made of large strands, each of them a bundle of smaller strands, the muscle fibers. A single muscle fiber may be centimeters in length, but is only about 50 micrometers (a millionth of a meter or a 25,000th of an inch) in diameter. A muscle fiber is a very long cell. Inside the cells are bundles of actin and myosin. Each myosin head, the active force-generating part of the whole assembly, is only 30–40 nanometers in size. Every gram of muscle contains about 10^{17} (100 million billion) myosin molecules.

The process by which myosin II generates motion is still not completely understood, but substantial progress has been made by structural (X-ray, electron microscopy), biochemical, fluorescence, and laser tweezer studies. The cycles of attachment and detachment, as well as ATP binding, hydrolysis, and release, are essentially the same for myosin II and myosin V. However, myosin II does not seem to be processive. It does not walk along actin filaments for long distances, as myosin V does. On the other hand, myosin II can produce faster motion than myosin V can.

As mentioned in Chapter 6, some people believe that myosin II is a weakly coupled motor, while others believe in a distinct, fixed-distance power stroke. Although myosin II was the first molecular motor discovered, it remains one of the most enigmatic. For example, as in myosin V, the role of ATP hydrolysis does not seem to be settled: In dozens of papers, most authors seem to reduce the function of ATP hydrolysis to an "increase in affinity to actin," while a few researchers assume that ATP hydrolysis is associated with some prepower stroke. Moreover, as in myosin V, the ATP hydrolysis proceeds in several steps: splitting off a phosphate to create ADP-P, and only then releasing the phosphate, which results in a bound ADP. These steps happen too fast for scientists to clearly link them to mechanical motions of the myosin. Solving this problem will require the development of faster fluorescence techniques. The hydrolysis of ATP releases a fair amount of energy, so it is difficult to imagine that it would simply serve to make the myosin head more attractive for actin. Perhaps it plays a role in biasing diffusion, for example, by internally changing the strain in the molecule.

How, then, is muscle motion controlled? One explanation is that some mechanism controls the supply of ATP to muscle. If we want to move, our bodies supply ATP; if not, they shut off the ATP supply. But this is not how it happens. As a matter of fact, in myosin II, the binding of ATP allows the myosin head to detach from actin. Therefore, a lack of ATP leads to rigor mortis: The myosin gets stuck on the actin and the muscles harden up. Instead, the motion of muscle is controlled by a molecular switch called tropomyosin. Tropomyosin binds to actin and blocks the site where myosin takes hold of the actin filament. When we think about lifting an arm, nerves send a signal to the muscle cells, which release calcium ions. The calcium ions bind to the tropomyosin and it releases the actin, thereby exposing a patch to which myosin II can bind.

THE MYOSIN SUPERFAMILY

Myosin, like kinesin, forms a large family with many functions (Figure 7.9). Muscle myosin II generates power in muscles, while nonmuscle myosin II helps cells change shape, move, and divide. Myosin V is a short-distance transporter in our cells. What about the other myosin molecules? Myosin I,

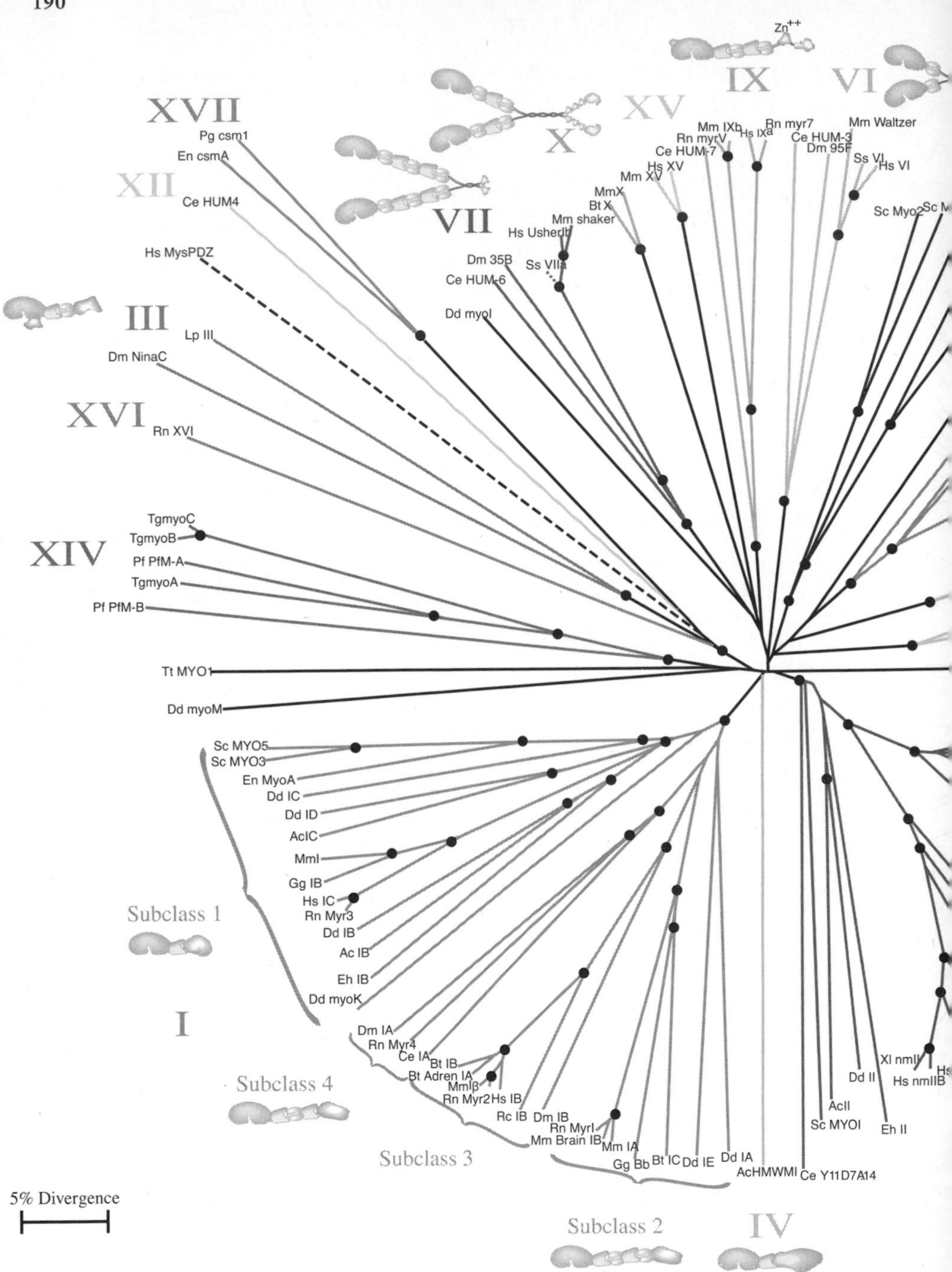

XVII
Pg csm1
En csmA
XII
Ce HUM4
Hs MysPDZ
III
Lp III
Dm NinaC
XVI
Rn XVI
XIV
TgmyoC
TgmyoB
Pf PfM-A
TgmyoA
Pf PfM-B
Tt MYO1
Dd myoM
X
VII
XV
IX
VI
Zn++
Mm IXb
Hs IXa
Rn myr7
Mm Waltzer
Rn myrV
Ce HUM-3
Ce HUM-7
Dm 95F
Hs XV
Ss VI
Hs VI
Mm XV
MmX
Bt X
Mm shaker
Hs Usherlb
Dm 35B
Ss VIIa
Ce HUM-6
Dd myoI
Sc Myo2
Sc MYO5
Sc MYO3
En MyoA
Dd IC
Dd ID
AcIC
MmI
Gg IB
Hs IC
Rn Myr3
Dd IB
Ac IB
Eh IB
Dd myoK
Subclass 1
I
Dm IA
Rn Myr4
Ce IA
Bt IB
Bt Adren IA
Mmlβ
Rn Myr2
Hs IB
Rc IB
Dm IB
Rn MyrI
Mm Brain IB
Mm IA
Gg Bb
Bt IC
Dd IE
Dd IA
AcHMWMI
Ce Y11D7A14
AcII
Sc MYOI
Dd II
Eh II
Xl nmII
Hs nmIIB
Subclass 4
Subclass 3
Subclass 2
IV
5% Divergence

A Myosin Family Tree

Tony Hodge and Jamie Cope

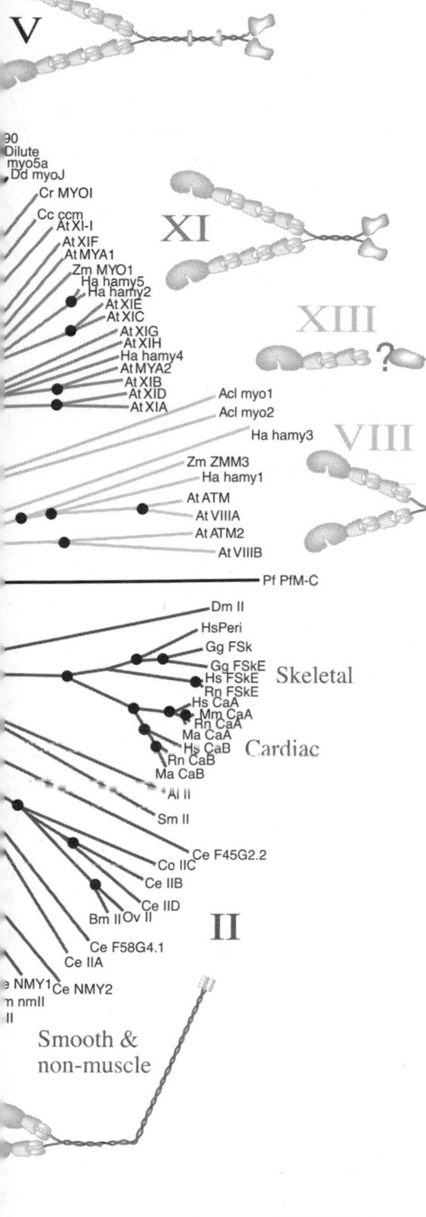

Abbeviations

Ac	*Acanthamoeba castellanii*
Acl	*Acetabularia cliftonii*
Ai	*Aequipecten irradians* (scallop)
At	*Arabidopsis thaliana* (thale cress)
Bm	*Brugia malayi*
Bt	*Bos taurus* (cow)
Cc	*Chara corallina*
Ce	*Caenorhabditis elegans*
Cr	*Chlamydomonas reinhardtii*
Dd	*Dictyostelium discoidium*
Dm	*Drosophila melanogaster*
En	*Emiricella nidulans (Aspergillus)*
Eh	*Entamoeba histolytica*
Gg	*Gallus gallus* (chicken)
Ha	*Helianthus annus* (sunflower)
Hs	*Homo sapiens* (human)
Lp	*Limulus polyphemus* (horseshoe crab)
Ma	*Mesocricetus auratus* (hamster)
Mm	*Mus musculus* (mouse)
Oc	*Oryctolagus cuniculus* (rabbit)
Ov	*Onchocerca volvulus* (a nematode)
Pf	*Plasmodium falciparum*
Pg	*Pyricularia grisea* (rice blast fungus)
Rc	*Rana catesbeiana* (bullfrog)
Rn	*Rattus norvegicus* (rat)
Sc	*Saccharomyces cerevisiae* (yeast)
Sm	*Schistosoma mansoni*
Ss	*Sus scrofa domestica* (domestic pig)
Tg	*Toxoplasma gondii*
Tt	*Tetrahymena thermophila*
Xl	*Xenopus laevis*
Zm	*Zea mays*
Adren	Bovine Adrenal (myosin I)
Bb	Brush Border Myosin I
CaA	Cardiac alpha (myosin II)
CaB	Cardiac beta (myosin II)
csm	Chitin synthase-myosin
FSk	Fast Skeletal (myosin II) – striated
FSkE	Embryonic Fast Skeletal (myosin II)
HMWMI	High Molecular Weight Myosin I
neur	Neuronal (myosin II)
nm	Non-muscle (myosin II)
PDZ	Human myosin with a PDZ domain.
Peri	Perinatal (myosin II)
sm	Smooth muscle (myosin II)

FIGURE 7.9. An example for the family tree of a molecular motor: the family of myosins. Reproduced/adapted with permission from Tony Hodge, M. Jamie, and T. V. Cope, "A Myosin Family Tree," *Journal of Cell Science* 113 (2000): 3353–3354.

unlike myosin II and V, is a *single-headed* motor used in cell motion. It attaches to the cell membrane on one end and to actin at the other. Then it pushes actin filaments around as they are polymerized (generated). In this way, it shapes the internal skeleton of the cell and provides tension between the skeleton and the cell membrane.

Myosin VI is a motor protein like myosin V, but it moves in the opposite direction of myosin V. While myosin V moves lipid-encased cargo to the outside of the cell, myosin VI moves cargo to the inside. It is involved in *endocytosis*, which is the process through which the cell ingests molecules from the outside of the membrane. Myosin VII seems to be involved in the cells in our ears: Mutations in the myosin VII gene can lead to deafness. Myosin IX may play a role in sending signals to the cell that initiates rearrangement of its actin filament network. Clearly, considering the large number of motor proteins and the multiple functions they fulfill, more research will be needed to figure out what they all do.

The Recharging Station: The Machinery of Energy Transduction

Through hydrolysis, molecular motors turn ATP into ADP. Then they release ADP and the spent molecular fuel pellet floats away. What happens to it? Wherever possible, cells are keen on recycling. Why not recharge the ADP by reattaching a phosphate and turning it back into ATP? Indeed, this is what the cell does, and it happens in one of the cell's most important organelles: mitochondria.

Mitochondria are the fuel recharging stations of cells (they are sometimes called the cell's power stations). They use energy derived from food to recharge ADP. The details took decades to decipher; the process is surprisingly complicated and proceeds through many steps. Fortunately for our story, at the end of the process, there is a marvelous molecular machine—a machine that, like no other, illustrates how, at the nanoscale, different types of energy can be converted into each other with very high efficiency.

But before we talk about this amazing machine, let's briefly look at how food turns into ATP-energy. As described in Chapter 1, scientists slowly came to the realization that the heat of the body was generated by a slow burning process. Experiments by Lavoisier, Liebig, Helmholtz, and others showed that the energy generated from food is roughly equivalent to the burning of food in air. But how do our bodies burn food? In our cells, a number of complicated, multistep biochemical reactions can turn various food molecules—sugars, proteins, fat—into standard, high-energy molecules used by the cell. In each biochemical step, some energy is removed in the form of ATP or other energy-carrying molecules, and the end product has correspondingly less energy than the original food molecules. Thus, breaking down the burning process into many steps makes burning a very slow and, consequently, very efficient process.

In one such burning process, glucose (a sugar) is turned into two molecules of pyruvate (a small organic molecule), while generating two ATP molecules. Prior to the availability of free oxygen on our planet (courtesy of photosynthesis), this reaction was the end of the road. You could get two ATP molecules out of one glucose molecule, and that was it. In anaerobic bacteria (bacteria that do not use oxygen), pyruvate is waste. But pyruvate still has energy to offer. In aerobic cells (like our own), pyruvate becomes feedstock for the next set of biochemical reactions. All in all, the two pyruvate molecules that result from the burning of one glucose molecule generate another thirty ATP molecules. Now, that's efficiency!

The first step in digesting pyruvate is to break it down further, releasing CO_2 (which we breathe out) and creating another intermediate product, an acetyl, which gets attached to a carrier called coenzyme A. This product, acetyl coenzyme A, then enters the next cycle, called the Krebs cycle after its discoverer, and after more CO_2 is released, we are left with several energy-carrying molecules called NADH and $FADH_2$. I'll spare you what these acronyms stand for—what's important is that NADH and $FADH_2$ are electron-rich molecules, which means they can easily give up electrons.

When NADH gives up electrons to an enzyme embedded in the membrane of the mitochondrion, NADH turns into NAD^+ (Figure 7.10). The split-off hydrogen ion (H^+, the same thing as a hydrogen nucleus, which in hydrogen's case, is a single proton) passes through the membrane-spanning enzyme to the other side of the membrane. The electron is passed through several carriers along a chain of more enzymes (called complexes I, III, and IV), which use the electron energy to pump more protons across the membrane. In the end, this *electron transfer chain* succeeds

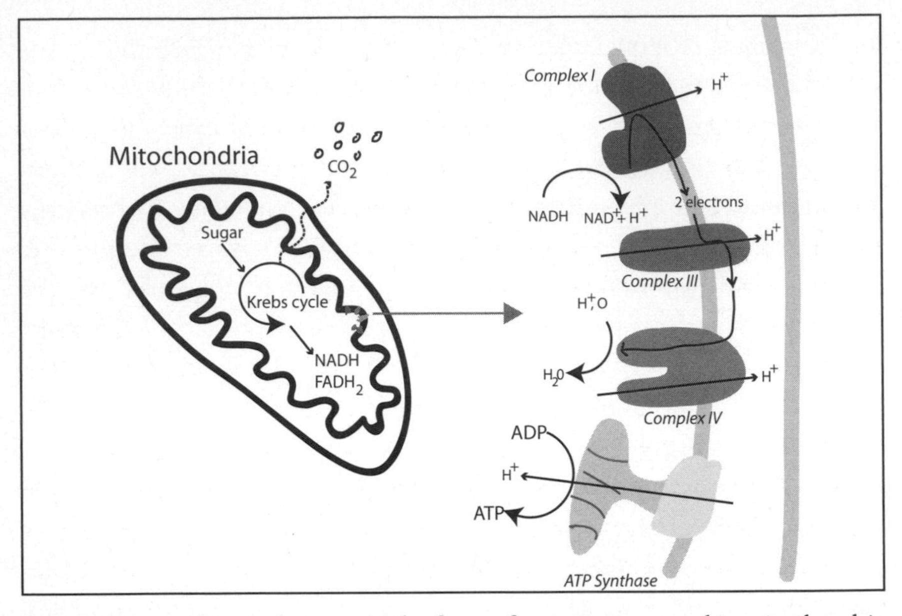

FIGURE 7.10. Chemical energy in the form of sugar is converted in mitochondria into energy in the form of ATP. This process takes many steps. First the energy is fixed in molecules called NADH and $FADH_2$. These molecules are then processed by different molecular machines in the mitochondrial membrane; the machines use the energy to pump hydrogen ions (H^+) from one side of the membrane to the other. NADH is processed by complex I, while $FADH_2$ is processed by complex II (not shown). The excess of hydrogen ions on one side of the membrane is then used to drive the ATP synthase, which recharges ADP to ATP.

in pumping a lot of positively charged protons from one side of the membrane to the other. This movement of charges (protons) leads to the development of a *voltage* across the membrane. Thus, the purpose of the electron transfer chain is to recharge a biological battery. The first energy conversion is complete—the cell has turned chemical energy into electrical energy.

The idea of storing energy as an excess of protons on one side of a membrane goes back to the Scottish biochemist Peter Mitchell, who in 1961 suggested this "chemi-osmotic" process as a key to understanding mitochondria. His ideas were subjected to much criticism. Most biochemists believed in a purely chemical process of recharging ADP. Many years were wasted searching for an enzyme that would chemically recharge ADP, but none was found. Finally, Mitchell's ideas were accepted and he was awarded the 1978 Nobel Prize for Chemistry. In the late 1970s, however, the details of how it all worked were still quite sketchy. How did the stored electrical energy lead to the recharging of ADP molecules?

ATP SYNTHASE AND THE AMAZING SPINNING BATON

Wayne State University hosts the Ahmed Zewail Gold Medal Award and lecture (named after the Nobel Prize–winning Egyptian scientist who pioneered ultrafast measurements of chemical reactions). In 2010, the recipient of the award was another Nobel laureate, Sir John E. Walker, who deciphered the structure of ATP synthase, the molecular machine that constitutes the last step in the slow-burning process in our bodies. I thoroughly enjoyed Walker's lecture, and during the question and answers, I asked about what was known about the evolution of these amazing machines. He raised an eyebrow and responded, in the laconic style of English intellectuals, "I usually only get questions like this in Texas," alluding to the creationists' use of the intricacies of our cells as proof of special creation. I laughed and assured him that I was genuinely interested in evolution and was not from Texas (my German accent should have been a giveaway).

Walker shared his Nobel Prize with UCLA biochemist Paul Boyer. Where Walker figured out the structure of ATP synthase, Boyer figured out how it worked. ATP synthase, as the name suggests, is a machine that makes ATP. The enzyme does not make it from scratch, but rather recycles ADP by attaching fresh phosphate groups. Attaching a phosphate to ADP and turning it into ATP requires energy, and ATP synthase takes this energy from the stored electrical energy provided by the electron transfer chain: The protons, laboriously moved to the outside of the mitochondrial membrane by the complexes of the transfer chain, are now pumped back—but not before giving up their electrical energy to the synthase machine.

In 1977, Boyer suggested that the flow of protons back through the membrane would drive a little rotary motor. Like the dial on a gumball machine, each turn of the motor would perform a different step of the ATP synthesis: binding an ADP and a phosphate, attaching the phosphate, releasing ATP. All of these steps would happen in the same enzymatic pocket. The rotation would somehow modify this pocket during each turn so that it would be most suitable for the particular reaction step it was assisting. Having three such multifunction pockets would allow three ATPs to be produced per full turn of the machine.

Boyer also suggested that the energy supplied from the protons moving across the membrane (the proton-motive force, as it was called) was not used to attach a phosphate to an ADP. This apparently happened readily in the pocket of the enzyme. Instead, energy was needed to *release* the newly formed ATP from the pocket. As Boyer and his group continued their research, using a special isotope of oxygen, ^{18}O, to follow the movement of molecules, they discovered a curious phenomenon: When they ran the process in reverse, letting the synthase break down ATP, rather than synthesizing it, they found that removing the resulting ADP from the surrounding solution stopped the reaction. This made no sense. In chemical reactions, the removal of the product will only speed up a reaction, since it creates a large imbalance between reactant and products. How could the removal of the product, ADP, stop the reaction? One of Boyer's students suggested that the still-unknown enzyme responsible for ATP synthesis (or breakdown, if run in reverse) only worked when somewhere in the enzyme, ADP was attached. This idea implied some kind of collaborative, allosteric interaction within the enzyme. Slowly, it became clear that the enzyme had three catalytic sites, which processed ADP in a sequential, coordinated way. The finding led Boyer to suggest that the ATP synthase went through a rotational cycle as it attached phosphates to ADP.

An interesting suggestion, but this theory only became widely accepted when Walker solved the structure of the F_1 part of the synthase. F_1 is the part Boyer identified with the dial—it performs the actual ATP synthesis. F_1 consisted of three identical units (each in turn consisting of two subunits, a and b) arranged in a circle like petals on a flower. Moreover, each identical b subunit had an ADP/ATP binding pocket. This circular arrangement was highly suggestive of rotary motion. But this motion was not proven until 1997, when a stunning experiment directly visualized the rotation of ATP synthase using fluorescence methods. Biochemist Hiroyuki Noji and his coworkers, then at Tokyo Institute of Technology, attached a short fluorescently labeled actin filament to the top of the F_1 unit of a single ATP synthase. When the researchers fed the synthase with

ATP, the machine ran in reverse, breaking down ATP into ADP and using the energy of ATP hydrolysis to fuel the rotation. This rotation then spun the attached actin filament around, like a majorette spinning a baton. Observing the actin filament in their microscope, Noji and colleagues saw that the F_1 unit rotated clockwise at a pretty good clip, up to four rotations per second. Moreover, the generated force was appreciable—by their estimates as much as 45 pN.

The efficiency with which the machine could turn energy from ATP hydrolysis into a rotation was astounding. In 2000, Noji's group found that the machine was at least 88 percent efficient. Such a high efficiency is unheard of for macroscopic machines, but it gives us an idea of how the human body can be so efficient overall. We are based on tiny nanomachines—and only a nanomachine can be this efficient!

Noji's experiment proved that the ATP synthase was a rotary machine, but the researchers had run the machine in the opposite direction of how it would be used in mitochondria. They had seen that a clockwise turn broke down ATP. It seemed obvious that a counterclockwise turn would achieve the opposite: reattaching phosphates to ADP. But this was just conjecture until experimentally proven. Unfortunately, it turned out to be difficult to observe ATP synthase in its usual environment in the mitochondria of a living cell. Once you remove the synthase from the cell, it is removed from the rest of converting machinery. Thus, the proton-motive force to drive the motor cannot be established and the motor cannot be driven in the direction that converts electrical into chemical energy.

Finally, in 2004, Noji's team found a way. Instead of letting the protons do the work, they attached a magnetic bead to the center of the F_1 unit of the synthase. Then, using a rotating magnetic field, they forced the magnetic handle either clockwise or counterclockwise. In other words, they hand-cranked the motor. When they cranked it clockwise, the amount of ATP in the surrounding solution decreased. The F_0 unit acted as an ATPase, splitting off phosphates from ATP and releasing ADP. This confirmed their previous experiments. However, when they cranked the unit counterclockwise, ATP increased. The motor was now making ATP out of ADP.

The F_1 unit is coupled to an F_0 unit, which is stuck in the membrane of the mitochondria. Unless cranked by some other motor, the F_1 unit will

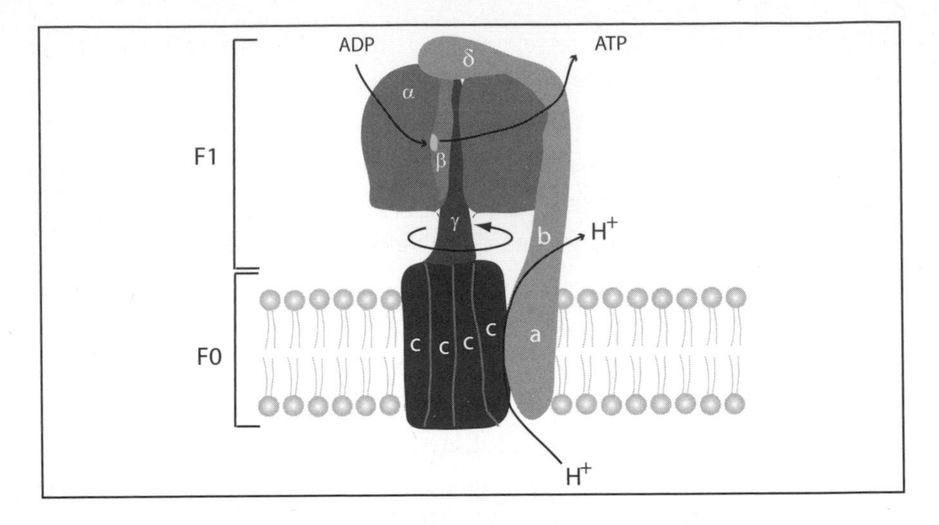

FIGURE 7.11. The ATP synthase molecular electromotor and recharging station consists of the F_0 electromotor, which is embedded in the mitochondrial membrane, and F_1, the rotary ADP recharging enzyme. F_0 consists of ten to fifteen c-subunits (depending on the organism) and an a-subunit and two b-subunits, which act as the stator. F_1 consists of three a-subunits and three b-subunits. The catalytic sites are on the b-subunit and are modified by the rotating shaft (γ-subunit) to perform the different steps of the ATP synthesis.

never synthesize ATP, it will only break it down. It needs to be coupled to a motor that cranks it in the counterclockwise direction—opposite the rotation during ATP breakdown. In our cells, this motor is contained in the F_0 subunit. The F_0 unit consists of several subunits, but overall, it consists of a ring of identical c-subunits, a large a-subunit, and two b-subunits, which reach up to hold the F_1 unit in place on top of the F_0. Attached to the c-subunit ring is a flexible shaft (subunit γ), which transmits the rotation of the F_0 unit to the F_1 unit (Figure 7.11). The γ-subunit is asymmetrical. As it rotates, it alternately pushes on the b-subunit catalytic domains and changes the shape of the catalytic pocket to initiate the different stages of ADP to ATP processing. In other words, the catalytic site is adjustable to either accept ATP binding, ATP hydrolysis, or ADP release. The rotating shaft deforms each unit in turn to perform these steps in a sequential manner (Figure 7.12).

The F_0 motor is a biological electromotor. The voltage created by the electron transfer chain forces protons to flow back across the membrane.

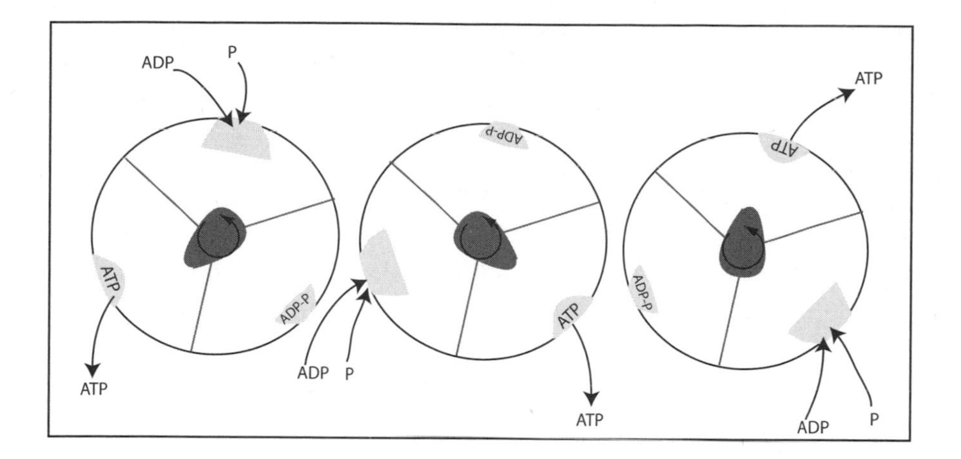

FIGURE 7.12. The F_1 unit of ATP synthase seen from above, as the γ-subunit, or shaft, rotates counterclockwise. The shaft deforms the binding pocket in the identical three b-subunits such that at different times, either ADP and P bind, ADP and P are combined to ATP, or ATP is released. For every 360-degree rotation, this machine produces three ATP molecules.

However, they first have to flow through the F_0 motor. The c-subunits of the motor have sites that accommodate protons. As protons enter the membrane, they are attracted to these sites and stick there. This site interacts with a hydrophobic patch in the membrane, which repels the proton. This creates a tilted energy landscape, and by a Brownian ratchet mechanism, the motor rotates. This brings another c-subunit into a position to load a proton, and the cycle continues. As the F_0 unit rotates, each proton makes an almost complete circle before it arrives at a place where it is released and continues its journey across the membrane. When this motion is coupled to the F_1 unit of the synthase, the F_1-ATPase is run in reverse and becomes an ATP *syn*thase, attaching phosphates to ADP.

GETTING AWAY WITH LESS

The F_0 subunit structure was even more difficult to decipher than the F_1 subunit structure. From early measurements, it was believed that four protons were required to rotate the machine 120 degrees and therefore produce one ATP molecule in the F_1 catalytic unit of the synthase. This would require the F_0 unit to contain 4×3, or 12 subunits (called c-subunits) that drive motion when protons pass through. However, studies of ATP synthases in different organisms (ATP synthases are ubiquitous in everything from plants to amoeba to humans) showed that there was a variable number of c-subunits, depending on the species. Plants appeared to have 14 c-subunits, cyanobacteria 13–15, certain other bacteria 11, and, finally, yeasts and the bacterium *E. coli* (and, presumably, humans) 10. Since it is believed that each subunit uses one proton to initiate a rotation, this variable number of c-subunits suggests that there are variable numbers of protons needed to synthesize one ATP molecule. For example, if 14 c-subunits are present in plants, this would suggest that 14/3, or 4.7, protons are needed to make one ATP, but this does not seem the case from experiments. In plants, only 4 protons are needed. At the other extreme, 10 c-subunits would translate into 3.33 protons per ATP molecule, suggesting an improvement in efficiency. But this has not been proven, either. Many mysteries remain.

Twisting: The DNA Machines

A detailed discussion of the entire machinery that reads, translates, repairs, duplicates, and maintains DNA would easily fill an entire book. But for the purpose of showing how molecular machines are used in almost everything our cells do, a brief overview will suffice. The DNA in our cells is rolled up—and then the rolls, in turn, are rolled up again—into compact DNA-protein structures called chromosomes. This packaging of DNA into chromosomes is performed by molecular machines.

DNA contains the information both to make proteins and to regulate the making of proteins. It does not contain the information of how to make a human—at least not directly. There is no gene for a toe or an eye. There are genes to make protein components of toes and eyes. The actual development of a human is a complicated process and involves the accurate timing of the synthesis of many proteins, their interaction, their regulation, and the movement of molecules by molecular motors. Many of these immensely complicated processes are still not well understood, although we now have a much better understanding than we did just twenty years ago. One of these well-understood processes is how a protein is made according to the instructions contained in DNA.

DNA is a double-helix molecule with a backbone made of alternating sugars (deoxyribose, which gives DNA its *D*) and phosphates. It looks like a twisted ladder. The rungs of the ladder are composed of various combinations of four nucleotide bases that constitute the DNA alphabet. The bases are adenine (A), guanine (G), thymine (T), and cytosine (C). Adenine is the same molecule that is part of adenosine triphosphate (ATP). The two sides of the double helix are complementary—this means that an A on one side always pairs up with a T on the other, and a G will always pair with a C. This specific pairing is achieved by hydrogen bonds, which are placed in such a way that, for example, T can form hydrogen bonds only with A, but not with G or C (Figure 7.13). Hydrogen bonds on their own are relatively weak, but the large number of hydrogen bonds fastening the two helices together gives great strength to DNA.

There are multiple advantages to DNA's double-helix structure with complementary base pairing. For one, the redundancy provided by the pairing of bases allows cells to repair DNA when it is damaged. Second is the molecule's ability to replicate itself, as James Watson and Francis Crick stated in their famous paper: "It has not escaped our notice that the specific

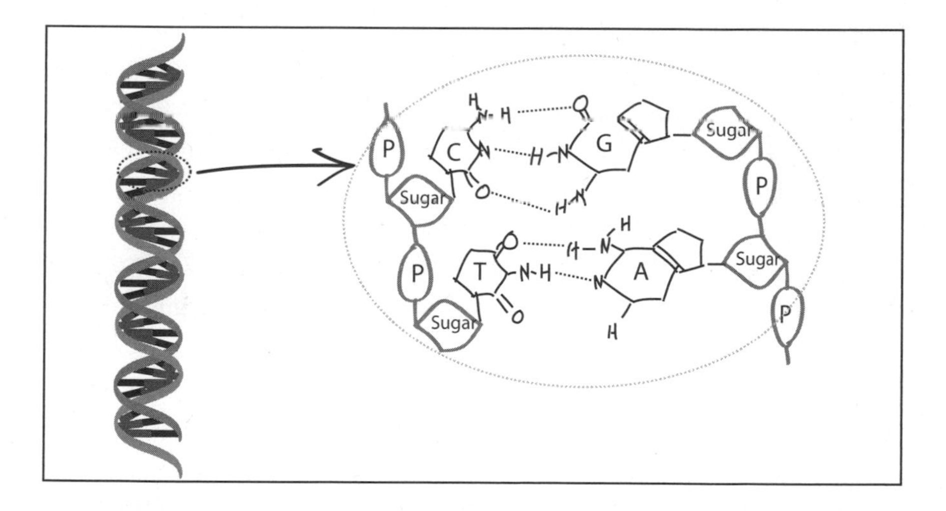

FIGURE 7.13. DNA structure. On the left is the overall double-helix structure. The two helices are made of alternating sugar and phosphate groups, while the rungs connecting the two helices are pairs of complementary nucleotides (A [adenine], C [cytosine], T [thiamine], or G [guanine]), as seen in the magnification on the right. The specific sequence of these nucleotide letters encodes the DNA message.

pairing we have postulated immediately suggests a possible copying mechanism for the genetic material." In other words, if you want to copy DNA, all you'd have to do is open up the helix and float some free bases around, and they will automatically pair up with their correct counterparts. Before long, you have made two helices out of one, both perfect copies of each other. It's not quite that easy, as we will see, but that's the basic idea.

The language of DNA uses three-letter words made up of four different letters (the bases A, C, T, and G). The three-letter word AAC, for example, codes for a specific amino acid in a protein. AAC happens to code for asparagine, while CUU codes for leucine. These three-letter words are called codons. The set of all three-letter words is the *genetic code*. A gene is a DNA sentence, a string of codons, which encodes a complete amino acid chain for a specific protein. For example, the KLC1 gene on chromosome 14 encodes the protein sequence for kinesin-1, our waddling motor protein.

The making of a protein proceeds through two distinct steps: transcription and translation. During transcription, the necessary information to make the protein is transcribed to another information carrier, RNA, which is a close relative of DNA. Why is this intermediate step needed? The DNA in our cells is kept safe in the cell nucleus. We do not want to cut out and remove parts of the DNA each time the cell manufactures a protein. Instead, the cell makes a temporary information carrier made of RNA, which can leave the cell nucleus and carry the information to where it is needed. This RNA information carrier is fittingly called messenger RNA (mRNA).

Thus, there are three major processes involving DNA in our cells: copying DNA when the cell divides (replication), copying DNA to RNA when a protein is to be made (transcription), and packaging DNA into chromosomes.

THE COPY MACHINERY

DNA is a very stable molecule. This is a good thing, as we do not want anything to happen to our DNA. Mistakes in our DNA can have disastrous consequences, such as the development of cancer. However, if the molecule is always curled up into a rigid structure, it would be impossible to copy or transcribe DNA. Therefore, we have to unravel, open, read, copy,

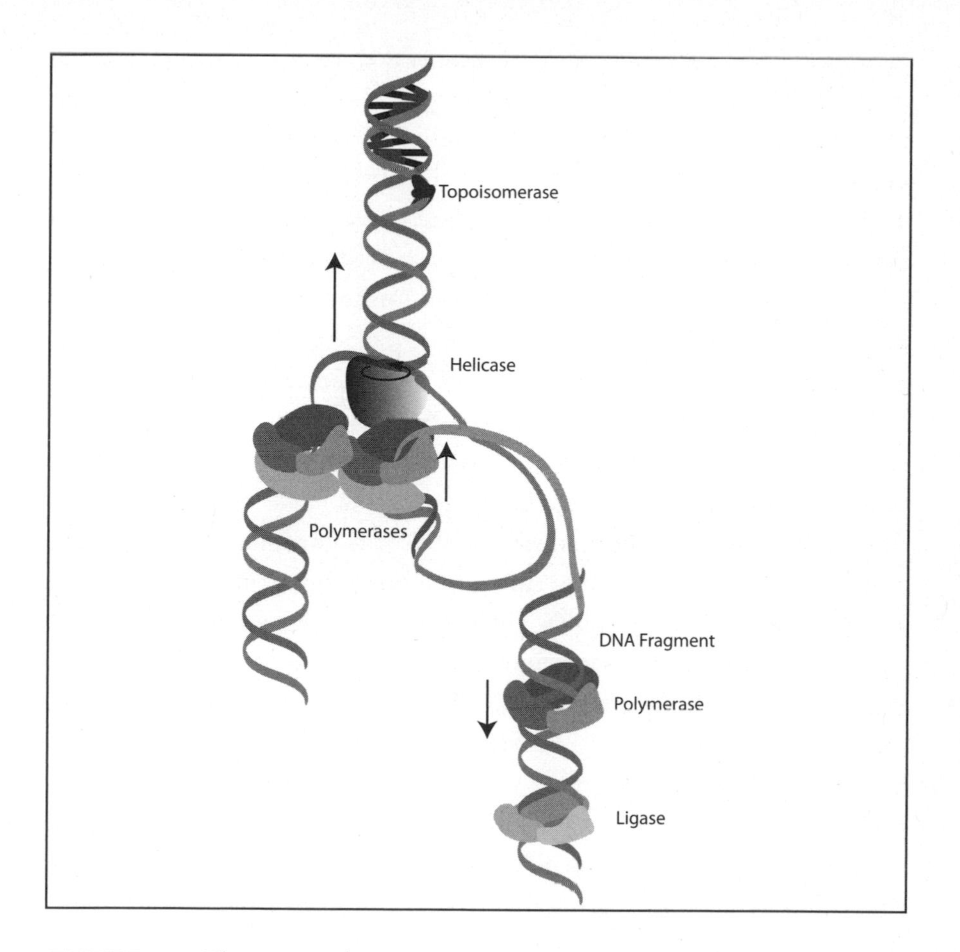

FIGURE 7.14. The DNA replication machinery. Helicase untwists the helix and separates the two strands. Topoisomerase cuts and re-splices the DNA to avoid tangles. Polymerases copy the DNA and make a new double helix out of each separate strand. One of the polymerases follows the helicase, while the other two work in the opposite direction. Thus they are forced to copy the DNA in fragments, which are spliced together by ligase.

and close it up again. DNA is copied when cells divide (each cell receives a complete copy of the genetic material), and it is transcribed each time a cell needs to make a protein. Although these processes seem superficially similar, they work in completely different ways.

During replication, a large number of molecular machines and enzymes work together to make sure the DNA is not damaged and that a true copy of the original is produced (Figure 7.14). But there are several complications. First of all, the DNA strand needs to be opened up. Second, if you have ever tried to untangle a twisted telephone cord, you can imagine that opening up a double helix can create a mess: Without proper precautions, the strands would quickly twist into a knot. To avoid a tangled mess, a molecular machine called topoisomerase snips one of the strands of DNA from time to time and passes it through the other to compensate for the final unwinding of the two strands by another molecular motor, helicase.

There is another complication: The two complementary strands of DNA are oriented in different directions. Like proteins (which have N- and C-terminals), DNA has different ends as well. The so-called 5' end has a phosphate group, and the 3' end has a sugar group. One of the strands runs from 3' to 5', while the other runs from 5' to 3'. Unfortunately, the machine that copies the DNA can only run in one direction: from 5' to 3'. This is fine for one of the two strands, because this copy machine, called DNA polymerase, can follow the machine that opens up the DNA (helicase). But the other strand points in the opposite direction. This means the copy machine on this strand has to move away from the point where the DNA is being split. What's worse, while this is happening, more DNA is split behind the machine. How to solve this problem? Copy a small section of the DNA, move the copy machine back, copy another small section, splice the sections together using another machine (called ligase), and so on.

TWISTING THE HELIX

Although space prevents us from discussing every molecular machine in DNA replication, we will concentrate on helicase, the machine involved in untwisting (and opening) the DNA helix. These untwisting machines have two jobs: separating the complementary DNA strands, and untwisting them to get them ready for copying. Untwisting the two separated strands can lead to more twisting in the not-yet separated DNA. Topoisomerases make sure the helicase motors do not turn the DNA into a tangled knot. Structurally, helicases look like a six-sided star with a hole in the middle through which DNA passes. How is the DNA spooled through the middle? Helicases self-assemble around the DNA, most likely by opening the six-sided helicase ring, letting the DNA in, and closing up again.

The six-sided shape may help the helicase continue to do its job while the DNA curls around as it moves through the ring center.

As discussed, the rungs of the DNA ladder—the complementary bases—are held together by hydrogen bonds. Studies have shown that helicases break these bonds much faster than these bonds would break on their own. Helicases are, first of all, enzymes that speed up the splitting of the DNA double helix. Remember, any bond will eventually break on its own. Every molecule is incessantly bombarded by the molecular storm—so there is always a possibility that, by chance, the bond will receive enough energy from the thermal chaos to break. The average time this takes depends on the height of the activation barrier. Helicases lower the activation barrier for splitting the two DNA strands by manipulating the charge on DNA and therefore speeding up the unzipping process. But helicases not only facilitate splitting the DNA; they also move along DNA like a pair of scissors. As they cut, they move forward, with the motion fueled by ATP. Helicases are therefore molecular motors as well. How this motion works is still being debated: Is it a tightly coupled power stroke or a Brownian ratchet mechanism? This debate may sound familiar. Some research suggests that ATP binding weakens the helicase's bond to DNA, allowing the helicase to move relatively freely. This would support a Brownian ratchet mechanism.

SENDING THE MESSAGE

The DNA in our cells contains all the information to make every protein needed by our bodies. To turn DNA instructions into actual proteins, the instructions have to be copied first to an RNA messenger, which transports the instructions to the factory floor, where the machines that translate DNA instructions into proteins are located. RNA is a more flexible and more ancient cousin of DNA (many people believe that life started with RNA). RNA contains a different sugar—ribose, as opposed to the deoxyribose in DNA—in its backbone, making it more flexible than DNA. Like DNA, RNA has four bases, three of which—A, C, and G—are the same as in DNA. However, the fourth base in RNA is uracil (U).

Because RNA also has four different bases, DNA language is easily transcribed into RNA language (simply replace the T with a U). This allows

RNA to play the role of messenger whenever the cell consults its DNA library to make a protein. The transcription of DNA into mRNA is accomplished by a molecular machine called RNA polymerase. This machine starts working when it encounters a specific DNA sequence, or *promoter*. The promoter marks the spot where the RNA polymerase is supposed to start transcription. Always the same, the promoter consists of two parts, ten and thirty-five base pairs before the sequence to be translated. Once the polymerase finds the promoter, it attaches to the DNA and starts making an RNA copy of the information it finds. RNA polymerase is the Swiss army knife of molecular machines. While DNA replication requires an army of enzymes and molecular machines, RNA polymerase works on its own—opening up DNA, transcribing it, moving along DNA, proofreading, and finally terminating the copying process.

The first step in transcription is to find the promoter. Unfortunately, DNA is impossibly long and coiled up. How can RNA polymerase find a specific sequence within the millions of sequences in DNA? Surprisingly, the polymerase finds its target very quickly, apparently by weakly attaching to the DNA and then sliding randomly along individual strands at a fast clip, helped by the molecular storm. Occasionally, the polymerase hops to make sure nothing gets missed. It is like shopping in a large store for a particular item: First you browse along the shelves in a linear fashion, but if it you can't find what you are looking for, you jump to a different area of the store and try again. Combining linear searching with jumps produces an efficient search strategy.

Once the promoter is found, the polymerase binds more strongly to the DNA. Now the motion becomes unidirectional. The polymerase starts transcribing, but only after a few wrong starts, which create junk RNA. To transcribe, the DNA strands need to be opened and unwound. RNA polymerase achieves this by creating a local loop of open DNA. This is different from replication, during which whole strands of DNA are separated and copied. The polymerase opens the DNA locally and quickly recloses it. Transcription is accomplished by matching each DNA base pair with a matching RNA base pair. Suitable base pairs are floating around in the form of triphosphates—such as ATP (which provides an A) and the corresponding GTP, CTP, and UTP. Using these supercharged nucleotides has a great advantage: The energized nucleotides not only supply bases to be inserted into the growing RNA strand, but also supply energy, which fuels the polymerase machine as it tracks along DNA.

Block and his coworkers, whom we briefly met when discussing kinesin motion, also measured the motion of RNA polymerase using their optical trap. They found that the polymerase moves in steps of one base pair at a time (0.34 nm). As the polymerase rides along the DNA, it drags behind itself a growing strand of transcribed RNA. Despite the drag force from the dangling RNA fragment, the polymerase machine is highly processive, transcribing thousands of base pairs without falling off the DNA. In one experiment, researchers applied 30 pN of force to the trailing RNA strand, but the machine was undeterred and continued on its merry way. This force is much bigger than would be needed to pull apart DNA and RNA (at the transcription site, they are temporarily bound through base-pairing). Therefore, RNA polymerase must serve as a strong clamp to hold RNA and DNA together.

MAKING PROTEINS

The third step in the DNA life cycle is the translation of the DNA information into an actual protein. This is achieved by the mother of all molecular machines: the *ribosome*. A human ribosome consists of about seventy-five proteins and multiple long and complicated RNA strands, all bound together into two major units, which come together to form the ribosome. Why RNA strands? So far, all the machines we have discussed were made of proteins. The reason RNA is so important in our cells is that it is versatile: RNA is an information carrier, like DNA, but because of RNA's greater flexibility, it can assume complex three-dimensional shapes, which can catalyze reactions like proteins. For this reason, many people believe that RNA came first—before DNA and proteins. This makes the ribosome, an RNA-based machine, especially interesting. Every living being on earth has ribosomes, and they all contain RNA. Every human cell contains about 100 million ribosomes, and tiny bacteria, like *E. coli*, contain about 10,000 of them.

The translation of genetic information into proteins requires a translator. The translation is handled by small molecules, which are part RNA and part amino acid and are called transfer RNA (tRNA). They work sort

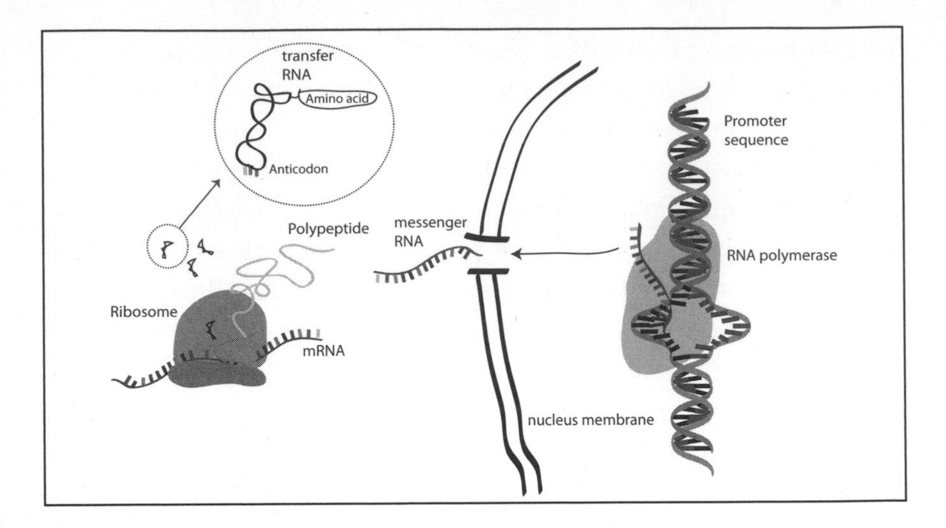

FIGURE 7.15. Transcription and translation: how DNA is translated into proteins. Right: The DNA message inside the cell nucleus is transcribed into messenger RNA (mRNA) by a machine called RNA polymerase. The mRNA leaves the nucleus. Left: The ribosome takes the mRNA and matches it up with a suitable transfer RNA (tRNA), such that the anticodon on the tRNA matches the codon on the mRNA. The amino acid on the other end of each tRNA is then attached to the growing polypeptide strand, which subsequently folds into a functional protein. The newly manufactured protein can be a structural protein like actin, an enzyme, or a molecular machine.

of like a dictionary: When you try to translate a word from English into German, you look up the English word in the left column and then find the German word in the right column. If the words are *flood* and *Überschwemmung*, we find that the two words have little in common. Without the dictionary, we would never guess that they mean the same thing (my wife teases me that German words are always three times longer than the English equivalent). tRNA works the same way: On one end of the tRNA molecule is the three-letter codon sequence, and on the other end is the corresponding amino acid. There is no way to know that AGU stands for serine, or GGC for glycine. We need this molecular dictionary. Figure 7.15 shows the transcription and translation sequence of DNA.

The dictionary is also coded in DNA. Our DNA contains information on how to make enzymes called tRNA synthetases, which, as the name suggests, synthesize tRNA. These enzymes have special pockets, such that only the appropriate codon matches with the appropriate amino acid. The tRNA synthetases always make sure that there is an ample supply of all combinations of tRNAs, so the ribosome can do its work. Using the tRNAs as types, the ribosome works like a typesetting machine. It takes the incoming message (in the form of mRNA, produced by the RNA polymerase) and then moves along the mRNA step-by-step and matches up each codon on mRNA with an appropriate tRNA. Once it finds a match, it clips off the amino acid attached to the tRNA, and attaches it to a growing strand of amino acids—the protein in the making.

Note that even though the ribosome is more complex than the other molecular machines we have encountered, almost all of them have one thing in common: They move along a track. Kinesin moved along microtubules, myosin along actin filaments, helicase and RNA polymerase along DNA, and the ribosome moves along mRNA. In most of these examples, the track is not permanently affected by the molecular machine. But some molecular machines literally eat the track they are moving on.

Eat While You Walk

A couple years back, I was hosting the well-known biochemist Gregory Goldberg as a colloquium speaker at Wayne State University. Greg has a wonderful, dry sense of humor. At the beginning of his talk, he postulated "Goldberg's law of biophysics": "A physicist may be converted to a biologist, but the reverse transformation is forbidden by nature." He was being generous: It is just as difficult for a physicist to understand biology as it is for a biologist to understand physics. Trying to enter a new field is seriously hard work. We speak different languages even when we talk about the same things.

Gregory Goldberg is a pioneer in the study of a particular class of large molecules, *matrix metalloproteinases*, or MMPs. MMPs are enzymes. The substrates for Goldberg's MMPs are the constituents of the extracellular matrix. One example is collagen, a protein that forms strong fibers that give our bodies shape. Without collagen, we would all live life as "The Blob," in a slimy puddle on the floor. Our cells are trapped like spiders in a tangled web of collagen and other extracellular matrix materials. To move through this web, cells have to somehow cut their way through.

This is achieved by certain MMPs, which break down collagen. Without MMPs, dead cells could not be replaced, or the cartilage in your joints could not be restored. But MMPs have a dark side. Cancer cells co-opt MMPs to allow the cells to spread, causing metastasis. Other MMPs are involved in forming new blood vessels that feed cancerous tumors. MMPs are therefore a target of cancer research.

Goldberg has studied MMPs for almost twenty years, together with other colleagues around the world (a group that has been referred to as the "MMP Mafia"—a prominent member of this "mafia" is my colleague Rafael Fridman, who got me interested in measuring MMPs with AFM). In 2004, Goldberg and his colleagues at Washington University made a startling new discovery. They were trying to understand how a particular MMP, called MMP-1, breaks down collagen and diffuses. The rate of diffusion is given by the diffusion coefficient, but to measure the diffusion coefficient of single molecules is tricky. Molecules are invisible in an ordinary microscope. But Goldberg's physics colleagues had just the right technique.

Fluorescence correlation spectroscopy (FCS) is used to measure diffusion in polymers, nanoparticles, and liquids in confinement. The technique is quite easy to understand: A laser is focused on the sample and causes special molecules, called fluorophores, to emit light at a wavelength that is distinct from the laser's. When there is no fluorophore in the laser focus, everything will be dark. When a fluorophore wanders into the laser light by random diffusion, there will be a tiny flash of light. Because the fluorophores randomly move into and out of the laser focus, the measured light will fluctuate. Knowing the size of the laser spot, scientists can determine the diffusion coefficient by analyzing the noisy signal coming from the wandering fluorophores. By attaching a fluorophore to a larger molecule, such as MMP, they can thus determine MMP's diffusion constant.

Much to their surprise, Goldberg and his colleagues at Washington University found that their data did not fit any simple diffusion model. Instead, the molecules appeared to move in a fixed direction as they continue along the collagen fiber. How could this be? Molecules are subject to random thermal motion, which has no preferred direction.

As we have seen, directional motion requires the degeneration of free energy into heat. But MMPs are enzymes—they do not hydrolyze ATP to move around. How could they move unidirectionally without violating the second law of thermodynamics? It turns out that MMPs get their energy by eating the track they are moving on. This is a good strategy: MMP-1 is supposed to break down collagen. If the enzyme simply randomly diffused, it would break the collagen strand at random locations. Instead, MMP breaks down collagen as it moves along, systematically creating large gaps in the collagen through which the cell can move.

Each break in the collagen presents a high activation barrier, keeping MMP from moving in that direction. It therefore has no choice but to move away from the break. It then catalyzes another break, moves away from it, and keeps going. Goldberg and colleagues called this the burnt-bridge mechanism. Imagine driving on a very long bridge. As you drive along, you throw an explosive out the back of your car. The way back is now obstructed by a large hole in the bridge. You have no choice but to drive away from the holes you create. Keep doing that, and you will have no choice but to travel in a fixed direction. This mechanism avoids violating the second law, because the breakdown of collagen supplies energy, which is used to direct the motion of the track-eating MMP.

Routing: Active Channels

From the examples we have discussed so far, it may seem that almost all molecular machines move along tracks (with the exception of the ATP synthase, which rotates). But many other molecular machines in our cells hardly move. Pumps are one example. Our cells are filled with water, proteins, small molecules such as sugars, and lots of ions. Cells need to control the amount of small molecules and ions in their interior—otherwise, they could be in great trouble.

When I was a child in Germany, we had a large cherry tree reaching up to the second-floor balcony of our house. In the summer, you could pick ripe red cherries from the balcony. After a rain, the cherries were often split open. The reason was osmosis. Cherry skins are permeable to water, but they do not let sugar escape. The excess of sugar inside the cherry creates a nonequilibrium situation. The sugar molecules want to dilute themselves with more water to match the sugar concentration outside the cherry. When it rains, the sugar molecules get their chance. While they cannot escape the cherry, the imbalance of sugar inside and outside the cherry produces a pressure that drives water inside the cherry. The cherry swells up and finally bursts open.

The same would happen to cells if they could not regulate the amount of sugar, protein, water, or ions in their interior. Cells have a rather impermeable cell membrane, but this membrane is pockmarked with a vast array of specialized pores. Many of the pores are passive: They let certain molecules through if the molecules happen to diffuse that way. But some are machines—active pumps that pump certain ions or molecules at the expense of ATP.

Usually, ions will move across a membrane until they reach equilibrium. Passive pores will let ions diffuse, but the ions will only diffuse until equilibrium is reached. To create a nonequilibrium concentration difference between the inside and the outside of the cells, active pumping is required. One such pump, the *sodium-potassium pump*, moves sodium out of the cell and potassium into the cell, maintaining the typically high-potassium, low-sodium environment found in most cells. These pumps are extremely important for the cell. Our cells typically expend one-third of their energy to run their sodium-potassium pumps. In nerve cells, the fraction of energy used for these pumps can be as high as two-thirds.

We have seen that the mechanists, from Democritus to Helmholtz, were right: Life is based on machines—on pumps and motors. What these scientists could not know is that these machines are molecular nanobots that work very differently from any machine they could have imagined. The nanomachines of life are the offspring of chaos. They are driven by chaos, and as we will see in the next chapter, they are, at least in part, designed by chaos.

8
The Watch and the Ribosome

I have called this principle, by which each slight variation, if useful, is preserved, by the term of Natural Selection.

—CHARLES DARWIN

What, as hath already been said, but to increase, beyond measure, our admiration of the skill, which had been employed in the formation of such a machine? Or shall it, instead of this, all at once turn us round to an opposite conclusion, viz. that no art or skill whatever has been concerned in the business, although all other evidences of art and skill remain as they were, and this last and supreme piece of art be now added to the rest? Can this be maintained without absurdity? Yet this is atheism.

—WILLIAM PALEY

WILLIAM PALEY (1743–1805) WAS AN EIGHTEENTH-century clergyman, theologian, and philosopher. His work was so greatly admired by fellow clergyman that the Church made him the sub-dean of St. Paul's Cathedral and the rector of Bishopwearmouth, providing ample income for a philosophizing priest. Paley's moral writing was surprisingly modern. He attacked slavery and the active maintenance of poverty by the rich. He even defended the right of the poor to steal, if this was necessary to feed themselves and their families. In his natural history, Paley was a reactionary. Writing at the end of the scientific revolution, he espoused a natural philosophy that harkened back to prerevolutionary times. Yet, his writing was so persuasive that even a young Darwin became one of his admirers. What pleased his fellow churchmen was that, in their eyes, Paley destroyed the mechanical philosophy. They agreed with him that it was absurd to believe that matter can organize itself. Moreover, they concurred that belief in the self-organization of matter was akin to atheism.

The most famous passage of Paley's writing is the beginning of his *Natural Theology*: "Suppose I had found a watch upon the ground, and it should be inquired how the watch happened to be in that place; we perceive . . . that its several parts are . . . put together for a purpose . . . that the watch must have had a maker: that there must have existed, . . . an artificer or artificers who formed it for [a] purpose."

The analogy of living organisms with watches or clocks has been so persuasive that radical agnostics like La Mettrie, Christian apologists like Paley, and even twentieth-century scientists like Schrödinger have all employed this analogy to support their completely divergent points of view. The mere observation that the superficial similarity of an organism with a watch or a clock could support atheism, Christianity, and scientific mysticism should raise suspicion. Is it a good analogy at all?

To answer this question, let us ask it in a different way: Are our bodies similar to the machines we design? The answer must clearly be no. As much as Descartes or La Mettrie were impressed by the pumps, pipes, and levers that make up our body, we now know that the human body is something that emerges from the interactions of *molecules*. Of course, there are macroscopic pumps and pipes and levers in our bodies, but on their own, these cannot explain what makes us alive. Bacteria are also alive, but they have no heart, lungs, or arms. We can build a machine that contains pumps and levers, but it would not find its own food and reproduce.

If life, then, is based on molecules, it will necessarily be subject to the randomizing power of atomic motion. Chance will play an important role. And indeed, if we look at the motion of molecular machines or the mechanism of evolution, chance is a constant companion of life. This is not the case for a watch. The last thing we want in a watch is for chance to play any role—chance is only detrimental.

Paley ridiculed chance: "What does chance ever do for us? In the human body, for instance, chance, i.e. the operation of causes without design, may produce a wen, a wart, a mole, a pimple, but never an eye. Amongst inanimate substances, a clod, a pebble, a liquid drop might be; but never was a watch, a telescope, an organized body of any kind, answering a valuable purpose by a complicated mechanism, the effect of chance. In no assignable instance hath such a thing existed without intention somewhere." What I find interesting about this view is that he relegates a liquid drop to chance, when a liquid drop is very much the result of necessity. A pebble is, of course, part chance, but also the result of the formation of atoms from subatomic particles, the creation of heavy elements in supermassive stars and supernovas, the crystallography of complex silicate minerals, and numerous complicated geological processes, from volcanism to erosion. A wart, on the other hand, is the result of a sophisticated, *evolved* machine—a virus. Paley could not have known these things, but it shows how such deeply felt beliefs fare in the light of modern knowledge.

A watch clearly has an artificer—because a watch is a brittle design. It performs one specific task, it only works if all of its parts work, and it cannot cope with any influence of chance or chaos. In short, a watch is not really that complicated. As a matter of fact—apologies to my watchmaker father—watches are Tinkertoys compared with even the smallest organelle of a cell. On the other hand, many natural entities, which even Paley would have excluded from being designed, are vastly complex: stars, planets, mountains, volcanoes, weather patterns, and, yes, even pebbles.

Paley was similarly misguided on the topic of reproduction. He mused about the possibility of one day finding that the watch could make another watch, "similar to itself." Would that not be proof that the creator was even more sophisticated than we thought? The watch's ability to reproduce would only add to the complexity of the watch and would therefore make it even *more* likely that an artificer had created it. But this reasoning is flawed: If a watch could make another watch, it clearly would not need a creator. A watch would simply be the result of another watch. And if the new watch, as Paley argues, were merely "similar" to the original one, then why couldn't the new watch be a little bit better than the old one? Interestingly, Paley introduces chance into the argument when he says that the new watch is similar, but not identical. How similar? What determines what is the same and what is different in the offspring? What if there were millions of such watches, reproducing and exchanging information about their construction, passing every improvement to the next generation—would these watches not improve over time?

Before you object that watches don't have babies, let me remind you that I am simply following Paley's argument. Obviously, a watch that could reproduce itself would stop being a watch. A watch is an artifact made for an *external* purpose—to tell the time. Once the watch starts reproducing itself, it acquires an *internal* purpose—efficient reproduction. If we still had an external agent who selects watches for how well they tell time and only lets good timepieces reproduce, they would, over the generations, become better and better watches. But in the absence of an external agent, the watches would stop just being watches, because efficient reproduction would become their new raison d'être. After some time, they would radiate into many different machines—based on which could reproduce the best.

Now let's go one step further. Nothing works in the absence of energy. A watch needs to be wound up to work. A living organism must eat. Thus, reproducing watches would need to take in energy. They would need to find ways to beat competing watches in the race to sequester enough energy to reproduce. This would require them to find new ways of making a living. Biologists call these roles *niches*. Before long, we would not recognize our watches anymore. Few would still tell time. Their wheels would now be used for digestion or locomotion. The hands and the dial would be used for attracting a suitable partner with which to exchange information. Maybe, a glow-in-the-dark hour hand would drive the opposite sex wild. This is evolution—this is life.

Evolution

How do molecules evolve? Despite the histrionic debates in various American school boards, the mechanism of evolution is, as we saw in Chapter 1, quite obvious when contemplated with an open mind. This observation prompted Darwin's supporter Thomas Huxley to lament his not having thought of it first. Molecules are subjected to the same natural selection that applies to the more macroscopic parts of an organism. As a matter of fact, the evolution of proteins is a good way to see how evolution works, because there is a direct relationship between the protein's amino acid sequence and the information encoding this sequence in DNA. Every molecular innovation—every new molecular machine that transports cargo a bit faster or makes fewer mistakes when transcribing DNA—will give an advantage to the organism it inhabits. Consequently, better molecular machinery will become more prevalent in a population. Or when conditions change, new machinery will emerge to deal with the changed conditions.

A famous example of the evolution of proteins was the 1975 discovery of a strain of *Flavobacterium* in a wastewater pond at a Japanese nylon factory. The bacteria in this pond had evolved to eat chemicals associated with nylon manufacturing—chemicals that do not exist in nature. On further investigation, researchers isolated three enzymes that had evolved inside these bacteria and that helped the bacteria break down nylon. None of these enzymes existed in *Flavobacterium* strains that were not raised in the nylon pond. How did the bacteria invent the new enzymes? Bacteria multiply very fast and exist in large numbers. Therefore, they can evolve very rapidly. In this case, the DNA replication machinery of a few *Flavobacterium* cells apparently made a mistake. The machinery read off a DNA sequence from the wrong starting point, leading to a so-called frame-shift mutation. It so happened that the resulting protein was helpful in breaking down nylon, which is helpful when you live in a pond full of the stuff. Is such serendipity really believable? Absolutely. Just consider that a human body contains 10^{14} (a hundred thousand billion) bacteria. The Japanese pond must have contained much more than that. A typical time for bacteria to multiply is twenty to sixty minutes. Assuming the slower time and assuming that the nylon factory was in production ten years, the bacteria would have gone through 87,600 generations of gazillions of bacteria. Considering this enormous number of bacteria and the many generations they pass through, a rather unlikely mutation now moves into the realm of definite possibilities. But we are not done yet: The first enzyme may not have been good at digesting nylon, but a bad nylon-digesting enzyme was certainly better than none at all. Once the bad nylon-digesting enzyme spread through the bacterial population, it evolved and improved rapidly.

The glacial, step-by-step, and somewhat unpredictable process of evolution makes it difficult for people to believe that this mechanism could have led to the sophisticated machinery of our cells. But as the above example shows, sometimes evolution happens in a few years. Remarkably, the lion's share of the history of life (almost three-quarters of life's history, or three billion years) consisted of the evolution of single-celled organisms. Multicellular organisms only appeared in the last billion years. Why did it take so long for multicellular life to appear? When we look at the complicated machinery of our cells, an answer suggests itself: It took billions of years of evolution to turn the first primitive enzymes into our modern sophisticated cellular machinery. Multicellular organisms became possible only when a minimum degree of efficiency and sophistication was reached. This view of life's early evolution is supported by the observation that on a fundamental level, all multicellular animals (and all plants) are the same. Humbling as it may sound, at the nanometer scale little distinguishes a human from a fungus. The basic cellular toolkit is the same. The complexity of this kit justifies the length of time it took for it to develop. Once the toolkit was in place, evolution was free to create ever more amazing multicellular creatures, from octopi to redwood forests. In some sense, the real mystery of life lies at the molecular scale. This is where all the real work of evolution was done. The rest is icing on the cake.

The fossil ancestors of our molecular machines are, for the most part, gone forever. Proteins do not keep for over three billion years, and bacteria with primitive machinery would have been eaten a long time ago. Even the so-called archaea microorganism, which have been found to be significantly different from bacteria, are not really archaic. In some sense, bacteria and archaea are more evolved than we are. After all, they had a lot more time, and they reproduced much faster. With this in mind, is there *anything* that can be done to determine how molecular machines may have evolved?

Trying to figure out the exact evolutionary steps leading to the ribosome or a kinesin molecular motor is like trying to solve a crime hundreds of years after it happened. Who was Jack the Ripper? It is impossible to tell. The trail has gone cold. Yet, using the few reports and other scant evidence that remains, we can make some plausible arguments about what kind of person he may have been. In much the same way, when it comes to the evolution of molecular machines, we have to look at the few remaining clues and try to come up with a plausible story. In this case, *plausible* means that the story matches the evidence and is in accordance with known physics and chemistry. Once a plausible story has been hypothesized, parts of it can be tested in the laboratory. If it passes these tests as well, we end up with a *likely* story, but we will never get a proven story. Jack will remain at large.

An instructive example is the evolution of the ribosome. In Chapter 7, I suggested that RNA is believed to be a more ancient molecule than either DNA or proteins. In the chicken-and-egg problem of what came first—DNA, which encodes protein, or proteins, which are needed to read the DNA—the answer is clearly neither. RNA contains information and can catalyze reactions. It is a kind of egg on feet, which can lay its own eggs. No chicken needed.

RNA's ability to catalyze reactions is a fairly recent discovery. In 1982, Tom Cech and coworkers at the University of Colorado–Boulder discovered that a certain RNA strand in a bacterium was able to splice parts of itself and reconnect the RNA strand, without any protein-based enzymes. This was the first indication that RNA could act as an enzyme. It took another ten years before Harry Noller and his group at the University of California–Santa Cruz demonstrated that the RNA in the ribosome also has catalytic properties. Over the years, it became clearer that *all* the hard work in the ribosome is done by RNA. When researchers removed the protein components, the RNA was still able to process messenger RNA and produce an amino acid chain, although at some loss to efficiency and fidelity.

The finding that the ribosome needed its RNA, but not its proteins, suggested that catalytic RNA may have been the basic constituent of early life. In a paper in 2010, researchers from the Weizman Institute of Science in Rehovot, Israel, and the European Molecular Biology Laboratory in Heidelberg, Germany, suggested that the RNA pocket where the amino acid chain is assembled is universal in all ribosomes and may constitute the original ribosome precursor. If RNA really was first, and it could catalyze its own evolution through splicing and reshaping, it may have eventually hit on a structure that could produce proteins. After that, proteins that assisted the primitive organism by being better catalysts than the RNA could have formed. This first protein-producing RNA may not have been able to create well-controlled protein products, but less control would have also led to more mutations and possibly faster evolution. Eventually, the combined forces of RNA and proteins invented DNA, and the modern cell was on its way. It is a likely story.

Another way to consider the evolution of molecular machines is to look at family relationships. In the previous chapter, we mentioned that kinesin and myosin share many similarities in their motor domain. This suggests that they may have evolved from a common ancestral protein. Myosin and kinesin share certain loops in their switch domains, which are associated with shape changes upon binding of ATP. Intriguingly, they also share these structures with so-called G proteins, which are not machines, but molecular switches. Molecular switches communicate chemical signals from the outside of the cell to the inside. In Chapter 6, we speculated that molecular motors may have evolved from enzymes that could change their shape when they bound a control molecule. This allosteric effect is what makes molecular switches work.

The details of how G proteins connect to kinesin and myosin is lost in the fog of billions of years of evolution. Nevertheless, that motor proteins would have evolved from molecular switches is very plausible, and the relationship with G proteins confirms this idea. The relationship between kinesin, myosin, and G proteins shows that in evolution, similar parts in different molecules can often serve different purposes. The evolution of a sophisticated machine like kinesin does not require that each part be invented from scratch or that all parts come into existence simultaneously. When it comes to evolution, almost anything goes.

Let us imagine two proteins A and B, encoded by certain genes in our DNA. Proteins A and B perform different functions. What if part of protein B could help make protein A work better, or what if the combination of parts from A and B were to create a new protein with a completely new function? No problem. Sometimes, whole protein sequences are translocated in our genome, either through copying errors or by viruses. This can lead to the combination of different proteins and the creation of an entirely new line of molecular machines. An example is the nylon-eating enzyme of *Flavobacterium*.

Ever since kinesin and myosin came into existence eons ago, they have evolved into many different forms, forming large superfamilies. We have seen an example for such a family tree in Chapter 7. The same is true for almost any molecular machine. Every protein is part of a family of related proteins whose jobs are often quite different, but which are clearly descendants of a common ancestor. Evolution never ends; it is ongoing. Once evolution discovers a new trick, such as a walking molecular motor, it soon creates many variants, all fulfilling specialized functions.

A common objection of creationists is that some biological structures are "irreducibly complex." What they mean is that a structure has many interdependent parts, so that if you remove just one, the whole thing could not work. For example, how could a car evolve? The engine could not evolve without already having a whole car in place. But the car could not evolve without an engine. All the parts of a car must be designed to fit together. No part can be left out. Thus, goes the argument, molecular machines must be *designed*, just as a car is designed. This is because (following their argument), a molecular machine is only functional when all the parts are in place. There can be no intermediate evolutionary steps. Every previous version of the molecular machine—without all the necessary parts in place—would have been utterly useless.

There are a number of problems with this superficially persuasive idea. First, as we have seen, structures are often put together from parts that previously served a completely different purpose. Take the car example. Clearly, different parts of the car can be developed independently of the whole car. An engine can drive a stationary machine. The Cardan shaft of today's automobiles, as we saw in Chapter 2, was invented for a water pump. Pistons come from air pumps. Gears from watches. And so on. There are countless examples of such versatility in evolution. The bones in our middle ear evolved from a jaw bone of an ancestral amphibian. The evolution of motor proteins from molecular switches, like G protein, is another example.

The second problem with the irreducibility argument is that incomplete structures are not as useless as one might think. Take the eye. Is an eye without a lens really useless? It sure beats no eye at all. Almost all intermediate stages of eyes exist in nature, from mere light-sensitive spots of some microorganisms to the sophisticated eyes of mammals. The same applies to motor proteins. We have seen that two-legged motor proteins can be processive. However, one-legged motor proteins can work as well, although with much less efficiency and highly reduced processivity, using a pure Brownian ratchet mechanism. Indeed, as we have mentioned, there are one-legged kinesins—although the jury is still out as to whether they can pair to form a two-legged kinesin. Nevertheless, at least theoretically, there is no physical reason why such a one-legged kinesin would not work. It may not be as good as kinesin-1, but it's better than no kinesin at all.

Many biologists consider evolutionary changes of DNA the most important events in the history of life. One proponent of this view is evolutionary biologist Richard Dawkins, famous for his idea of the "selfish gene." While there is, of course, much to be said in favor of this view, biologists often underestimate the role of physical law. The DNA-centered view therefore emphasizes chance over necessity. This has been exploited by creationists who like to abuse the concept of chance in evolution to claim that evolution is random and that randomness alone could not have created the complexity of life.

If evolution were truly random, the probability of creating just one functional protein would be astronomically small. Calculating such probabilities is a common parlor game among creationists. But these probabilities are irrelevant. Evolution is *not* random: It is the collaboration between a random process (mutation) and a nonrandom, necessary process (selection). It is the result of the balance of chance and necessity. This is not unusual—*all of nature is the result of this balance*. If not, nature would be either a featureless structure that is the same everywhere (if necessity wins) or a random "mush" with no structure at all (if chance wins). The exquisite order and the amazing variety we see in nature at every level—from galaxies to molecules—is the result of the fruitful interaction of chance and necessity. What is the probability of Earth or a pebble? It's a meaningless question. Similarly, the question about the random assembly of a protein is also meaningless. Evolution is not random.

Another favorite question of creationists is "How did all the information get into DNA?" At first glance, questions about information in DNA seem legitimate. This is because the message in DNA has meaning. It encodes the structure of a protein or regulates the development of an organism. But is meaning the same as information? Information is measurable; meaning is not. Creationists conflate these two terms to suit their own ends. In information theory, a message has more information the more random it is. As we mentioned before, a perfectly ordered message contains little information. AAAAAAA contains no information, while ACTTGATTC contains information. But does ACTTGATTC have meaning?

The whole idea of DNA containing information is, in my opinion, one of the main culprits in maintaining the myth of creationism and intelligent design. First of all, without the genetic code and the entire machinery of transcription and translation, DNA contains neither information nor meaning. Worse, strictly speaking, DNA does not even encode proteins—at least not functional proteins. DNA only encodes the amino acid sequence of a protein. The functionality of a protein comes from its three-dimensional shape and the physical properties of various parts of the protein. This shape is the result of protein folding, which is the result of *physical forces* (sometimes helped by chaperonins) acting on a sequence of amino acids. Much of the information to shape a protein into its functional form is contained in the laws of physics and the action of these laws in space and in time. How would you quantify the information input life receives from physics and space-time? You can't. DNA only makes sense in the context of physical law, already-established order, and interactions with the environment. DNA does not tell us the final shape of an animal or a protein. Without context, DNA is meaningless.

An important statement of genetics is the *central dogma*: Information flows in only one direction, from DNA to RNA to proteins, never back from proteins to DNA. While the central dogma holds during replication, transcription, and translation, during the development of an organism, proteins control which parts of DNA are read at any stage of the development. There are feedback loops. The information to make a human being is therefore not encoded in DNA as in a blueprint. Although the word *blueprint* is often used for DNA, this is a misleading analogy. A much better analogy is *recipe*. To make a human being, DNA contains information to make proteins, which by their *physical interactions* with DNA, RNA, or other proteins, in the form of complicated regulatory feedback loops, shape the developing organism. This is similar to cooking a meal. A recipe does not contain a complete description of the result of cooking a meal; it just contains information about the ingredients (proteins) and the timing of adding the ingredients (regulation). Then the physical interactions between the ingredients take care of making the meal.

Another way to put this is that organisms are emergent phenomena, emerging from complex interactions according to a specifically timed recipe. There is no way you could completely specify, in genetic code, every cell in a human being. How would you specify the trillions of connections in our brain? A few years ago, it came as a bit of a shock when the Human Genome Project revealed there are only about twenty-three thousand protein-encoding genes in the human genome. This is not much more than the number found in simple worms. I think the utter insufficiency of the information in DNA to specify an organism is one of the most powerful arguments for evolution. As argued before, life is a complex game played on the chessboard of physics and chemistry. I can think of no better analogy. Development of an organism needs information about proteins, but also needs space, time, physics, and complex feedback loops. None of these are encoded in DNA.

Evolution, like life, is also a game on the chessboard of time, space, and physics. The outcome of this game cannot be determined a priori. The game can create an enormous number of possible outcomes, and the role of evolution is to find some of these outcomes. By tweaking a protein here, or regulating a DNA sequence there, we see what effect this has; evolution has created a world inhabited by a limited set of all possible creatures that could theoretically exist. One of them happens to be us.

Creationists argue that humankind is the goal of Earth's history. If you start with this assumption, it would certainly be difficult to see how a playful process such as evolution could have necessarily ended up with us. Once you abandon this idea, however, and realize that evolution would have come up with *some* viable organisms, but not necessarily the same we encounter on our planet, it all starts making sense. Am I arguing for chance here? Chance is important, but I believe that life is inevitable and that myriad forms of life would have evolved in any case. As we have seen before, pure chance creates chaos; pure necessity, rigidity. Chance and necessity together become creation. What is created may be unpredictable, but creation itself is unavoidable.

This discussion is reminiscent of the differences between the views espoused by D'Arcy Thompson and Jacques Monod (Chapter 2). Monod believed that the existence of life is an incredible accident, the winnings of a cosmic lottery. It seems that Monod was too caught up in the DNA-centered view of life. Thompson, on the other hand, lived before we even knew about DNA. He emphasized necessity—he believed that all structures in living beings are the result of mathematical and physical laws. This is also clearly incorrect. Physical law by itself can make a rock, but without information, provided by evolution, we cannot make a living being. The views of Monod and Thompson can be combined to arrive at a fuller and more creationist-proof view of life: Information is important, but information comes from many sources—evolution, physics, chemistry, and the interaction of many complex entities in living cells.

Ratchets

The interaction of chance and necessity in evolution is mirrored by the interaction of chance (as molecular storm) and necessity (structure and physical laws) in the functioning of molecular machines. The second law of thermodynamics predicts that everything moves toward bland uniformity. Yet we have seen that the emergence of the bewildering complexity around us does not violate the second law, as long as we pay the free-energy cost. Still, to arrive at this complexity, we need some kind of free-energy-fueled mechanism. In our cells, directed motion, "purposeful" activity, is created by the action of molecular ratchets—molecular machines, enzymes, and motors, which by degrading free energy and due to their asymmetric structures, can rectify the random motions of the molecular storm to create order. Evolution is also a ratchet: It rectifies the random input from mutations into the creation of an ever larger number of possible creatures. This rectification is achieved by natural selection. Thus there is a pleasing analogy between evolution and its products, our molecular machines.

There is also a more direct connection between the molecular storm and evolution. As we saw from Delbrück's green pamphlet, which inspired Schrödinger to write his book *What Is Life?*, thermal motion is the main contributor to mutations. Even more to the point, replication and DNA repair are performed by molecular machines, which are subject to the molecular storm and therefore sometimes, although rarely, make mistakes. These mistakes supply fodder for the ongoing evolution of life on the planet. Interestingly, evolution strives to minimize mutations. The extremely high fidelity of replication (one base-pair mistake in ten billion base pairs) shows that there is, paradoxically, an evolutionary advantage to not evolve. Evolution is rarely radical. The low error rate ensures that it is a gradual process. Nevertheless, small differences can sometimes have a large impact on the final result, because DNA encodes a recipe, rather than a blueprint. The same is true in cooking. For example, consider leaving the baking powder out of your cake!

The observation that evolution acts like a ratchet also discredits the probability arguments of the creationists. Evolution builds improbability step by step, mutation by mutation, selection by selection. The question "What is the probability of creating a kinesin by randomly combining amino acids?" is irrelevant to how evolution works. Kinesin did not spring into existence fully formed; nor was it a goal of evolution. It is simply something evolution stumbled upon, as it ratcheted up more and more complexity, one small change at a time.

There Is No Other Way . . .

Looking at molecular machines has made me realize that evolution is the *only* way these machines could have come to exist. As we have seen, life exploits all aspects of the physical world to the fullest: time and space, random thermal motion, the chemistry of carbon, chemical bonding, the properties of water. Designed machines are different. They are often based on a limited set of physical properties and are designed to resist any extraneous influences. The tendency of molecular machines to *use* chaos, rather than resist it, provides a strong case for evolution. Why? If life started by itself, without a miracle, then life had to start at the molecular scale. The molecular scale has always been dominated by the molecular storm. The ability of life to somehow incorporate thermal randomness as an integral part of how it works—as opposed to giving in to the chaos—shows that life is a bottom-up process. It is not designed from the top down. A top-down design would have avoided the complications of thermal motion by making the fundamental entities of life larger, so they could resist the molecular storm more easily. This is what machines designed by humans do—until recently, as nanotechnologists have learned from life's nanobots to create tiny machines of their own.

Molecular machines' exquisite adaptation to their molecular environment is also a strong argument for evolution. Evolution is tinkering—the gradual improvement and better adaptation of biological structures. The history of life has been long, and evolution had ample time to create these amazing physics-exploiting machines that run our bodies. To achieve such near perfection, you need a process that designs dynamically. A onetime design is not enough. Conditions change over time, and our molecular machines need to remain adaptable. An external designer would do best if the designer used evolution to do the work. Adaptation is assisted by the fact that physical laws provide the missing ingredient. For example, many structures in our cells are made through self-assembly processes, which are the result of physical forces (vesicles, collagen, etc.). Evolution does not overdesign: It designs just enough to take advantage of physical laws. If physics does the work for you, then why bother designing what is already designed?

We also ought to consider the commonality of the molecular apparatus in the cells of every living being. Many molecules and cellular processes in an *E. coli* bacterium, a yeast, a bluebird, a begonia, or a human are almost identical. This strongly suggests common ancestry. At the same time, looking at the differences between organisms, we see how various molecular machines have been adapted to fulfill specialized functions peculiar to each species.

Yet, the best argument in favor of evolution and against a static-design view may be that any designer would have to work hard to keep organisms from evolving. This is what I alluded to in the beginning of the chapter. Paley's reproducing watch would evolve. As we have seen, mutations happen. Some of these may impart an advantage to its bearer. Such an advantage would tend to spread through a population. How could you stop it? And why should anybody want to? If I were the all-powerful being in charge of the world, I wouldn't bother. Why not sit back, relax, and enjoy what wonderful things evolution can create for you?

9

Making a Living

Neither DNA, nor any other kind of
molecule can, by itself, explain life.

—Lynn Margulis, *What Is Life?*

THE EMINENT HARVARD BIOLOGIST AND WRITER ERNST Mayr (1904–2005) wrote in *This Is Biology* that when scientists and philosophers have talked about life, they often considered life as opposed to the lifelessness of "an inanimate object." The problem with this definition of life, according to Mayr, is that life seems to refer to some "thing"—an idea that has misled philosophers and biologists for centuries. If life is a thing, then it must be clearly distinguished from other things, and therefore the existence of a "life substance" or "vital force" must be invoked. However, Mayr explained, once we realize that life is not a thing, but a *process*, we can begin to scientifically study the process of living. We can make a distinction between living and nonliving. We could even attempt to explain how life's processes can be the result of molecules.

Nobody can explain what life is. This has always been the problem with the question "What is life?"—a question that has led philosophers and scientists to look for a life force not only for centuries, but for millennia. While we cannot define life, we *can* explain how life works. We can explain the process. The molecular biophysics and nanoscience revolution has succeeded in explaining, as Mayr puts it, "how living, as a process, can be the product of molecules who themselves are not living." This is an important step. To delineate life from the "lifelessness of an inanimate object," we need to first understand how molecules can generate directed motion and activity, that is, how chaos becomes order. The new science of molecular machines has been successful in doing just that. But is that enough? Is "living" the sum total activity of all the molecular machines in our bodies?

Unfortunately, understanding kinesin or ATP synthase does not explain human life or even that of a single cell. I can throw all the molecules of a living cell into a test tube and shake them up—some of the motor proteins may wiggle for a while—but they will not assemble themselves into a living cell. Are we then back to square one? Do we require some invisible force, after all, to coordinate the molecular activity in our cells? No, we should have learned enough to see that this is not necessary. What distinguishes living organisms is not that they exist outside physics, but is that they are based on a self-organized, dynamic structure that perpetuates the organization of the organism from one point in time to the next. Life sustains itself. Life comes from life.

Every person's atoms are replaced within seven years, yet we remain the same person. We are not the atoms that constitute us; nor are we our proteins, DNA, or molecular machines. We are, instead, a complex process, a program, as it were, running on chemomechanical hardware. The analogy of life with a computer program fits our modern times, where computers have taken over the iconic status once reserved for clocks or steam engines. However, we should be careful not to overuse the computing analogy of life. The "program" that constitutes the process of living is massively parallel, decentralized, self-adaptive, "squishy," and controlled almost entirely by exchanges of matter (with the exception of nerve impulses, but even there, some matter exchanges are involved). It is also a program that has evolved over billions of years.

Living is programmed molecular dancing. We cannot allow molecular motors to move random cargo to random places. Specific cargo needs to go to specific places at specific times. The same is true for every activity of our cells: What proteins should be produced? When? How many? When should a molecular motor take on cargo, and when should the cargo be released? In Chapter 6, we glimpsed how cells regulate such decisions. Most complex proteins can be controlled through allostery—the change in structure and activity when the protein binds a control molecule. In our cells, proteins regulate other proteins, but also the transcription of DNA. In turn, proteins are made by the instructions contained in DNA and are controlled by other molecules, including sugars, ions, and lipids. The complicated program of "living" emerges from complicated feedback loops between all these molecules, linking them together in complex networks. The idea of a dynamic state of complex feedback loops is difficult to fathom. But that is what is going on in a cell. The complex molecules of our cells are marvels of evolutionary engineering. But the cell only becomes a cell when these molecules cooperate in a rich network of regulated interactions. This cooperative, self-sustaining, regulated activity is what we call *living*.

Regulation

How does regulation work? When we talked about how molecular machines work, we never mentioned how they know when to do their work. Do they work all the time, grabbing cargo, moving it, pumping ions in and out of the cell, or producing more ATP, even if it is not needed? That would be a recipe for disaster. Molecular machines are enzymes, first and foremost, and most enzymes in our bodies are regulated by inhibitory binding or allosteric interaction. Inhibitory binding is a direct way to regulate an enzyme or a motor. If a molecule other than the enzyme's substrate binds to the enzyme's binding pocket, and this molecule cannot be transformed by the enzyme, then this molecule "gums up" the enzyme's function. This is inhibition. Allostery, as we have already learned, is the control of an enzyme's activity by the binding of a control molecule to a separate binding pocket on the enzyme, which then through some conformational change controls the binding of the substrate, either enhancing or inhibiting it.

Inhibitory binding is how most drugs (and many poisons) work. The antacid pantoprazole is a proton-pump inhibitor. The particular proton pump that this drug inhibits is a molecular machine in the membranes of the cells that line the stomach. The machine's "job" is to take energy from ATP and use it to pump protons (hydrogen ions) into the stomach. Protons make the stomach acidic. Pantoprazole binds to the machine and blocks it temporarily, inhibiting an increase of acidity in the stomach. Inhibitor binding strength, or how long the inhibitor will remain bound before it again frees up the enzyme, depends on the height of the activation barrier. This allows nature to tailor inhibitors to control reactions over different time scales. The same is true of allostery: The binding of a control molecule is controlled by the binding strength (the *affinity*), the rate at which the molecule dissociates (which depends on the activation barrier), and the concentration of the control molecule in the cell (at low concentrations, the probability that an enzyme and a control molecule will meet is low).

Kinesins are a good example of how molecular machines are regulated in the body. Kinesins consist of one or two similar motor domains, which process ATP and bind to microtubules. In addition to the motor domain, kinesins can contain a number of other domains to bind cargo, to bind to specific locations, or for regulation. In the cargo-carrying motor kinesin-1 (as well as in kinesin-2, kinesin-3, and kinesin-7), the cargo-binding domain also serves a regulatory function, as it would be wasteful to have motor proteins running around in the absence of cargo. After all, they use up ATP. How is kinesin regulated? When no cargo is around, the kinesin molecule folds up, such that the cargo-binding domain can bind loosely to the motor domains. In effect, the molecule puts on its parking brake. If the cell wants to activate the motor, it sends two control molecules, which bind to the cargo domain and release it from the motor domains. The process essentially takes off the parking brake. This is not the only way kinesin-1 is regulated. For instance, the cell can control which microtubule a kinesin walks on using various proteins that bind to the microtubule. Some will "attract" kinesin, and some will inhibit its motion on the track. Another set of regulatory proteins controls the binding and release of cargo. It is quite typical for enzymes and molecular machines to be regulated in various ways to make sure they do the right job at the right time in the right location.

The very existence of molecular motors is also regulated by various molecular feedback loops and control molecules. For example, certain different kinesins are needed only during a specific phase of cell division (mitosis). In this phase, these kinesins are manufactured at a higher rate. Once the next phase begins, other proteins direct the breakdown of these kinesins, which are then recycled. Mitosis is a complicated, highly choreographed process. Not only the presence of kinesins (which help to separate the chromosomes), but also their location must be regulated. Control proteins make sure the various kinesins do their work at the right place.

Understanding the roles and regulation of molecular machines has been a boon for the pharmaceutical industry. Medical drugs, with few exceptions, work by inhibiting enzymes or molecular machines and are artificial control molecules. The task of R&D personnel at major drug companies is to come up with chemicals that specifically bind to target proteins, blocking their activity, and not binding to anything else, as this would cause side effects. Drugs need to be specific.

Kinesins are a target for cancer drugs. Cancer cells divide prodigiously, creating tumors. Eventually, the cells spread, causing metastasis, the main cause of cancer deaths. A drug called monastrol targets kinesin-5, which plays a major role during mitosis. As described in Chapter 7, kinesin-5 is a double motor that can bind to two microtubules at the same time. Kinesin-5 controls tension in the spindle, which separates chromosomes during the cell division. When monastrol binds to kinesin-5, the drug causes a change in structure of its ATP-binding site. The kinesin can still bind ATP, but it can no longer release ADP after hydrolysis. With ADP stuck in its ATPase pocket, the motor cannot obtain energy and falls dead. The spindle falls slack, and the cell cannot divide. The cancer cell, stuck in the middle of dividing, commits cell suicide.

Side Effects

At Wayne State University, the lab of my colleague Rafi Fridman studies the interaction of the collagen-eating, membrane-anchored enzyme, MMP-14 (metalloproteinase 14), and its inhibitor TIMP-2. My lab collaborates with Rafi, trying to measure the affinity of TIMP-2 for MMP-14 on living cells. This requires measurements at the single-molecule level.

To do this, we attach TIMP-2 to an AFM tip and let it interact with MMP-14 on the surface of a living cell. Then we move the lever up to pull on the bond between MMP-14 and TIMP-2 and record the force needed to break the bond. After countless measurements and applying the proper statistics, we can determine the average lifetime of the bond between the two proteins.

The interaction of MMP-14 and TIMP-2 is of special interest because scientists previously discovered that TIMP-2 not only inhibits MMP-14, but also primes the enzyme to *activate* another, free-floating MMP, called MMP-2. In other words, the so-called inhibitor of MMP-14 inhibits the collagen-destroying activity of MMP-14, but at the same time *activates* another collagen-destroying machine, MMP-2. The inhibitor is not really an inhibitor, but rather switches from one method of destroying collagen to another. Why? This is a question of ongoing research.

As mentioned in Chapter 7, MMPs play an important role in the motion of cells. In cancer cells, MMPs are often produced in high numbers and allow cancer cells to spread throughout the body. Because of this implication for cancer, MMPs have been a major target for drug development. Like monastrol, drugs that target MMPs are artificial inhibitors that stop MMP from doing its work. When the first artificial inhibitors were developed, they worked very well—in a test tube. However, after lengthy approval processes, these drugs were used on terminally ill cancer patients, but did not work very well. Moreover, they caused serious side effects, especially excruciating joint pain. What happened? MMPs are not only used by cancer cells, but also by regular cells, including those that maintain the cartilage in our joints. Indiscriminately shutting down the activity of an important molecular machine is not the best way to battle cancer.

Incidentally, the kinesin-5-targeting drug, monastrol, also ran into trouble. Besides causing a number of side effects (noncancer cells like to divide, too), it was ineffective in many types of cancer. It is not always clear why certain drugs do not work as hoped from laboratory experiments. Shutting down a single type of enzyme or molecular motor may not always be the key to finding a cure for cancer or other diseases. The molecules in our cells interact with each other in complicated ways that we are just beginning to understand. Our cells are well-regulated machines. Deciphering their regulation has proved difficult, as their complexity is staggering. This lack of understanding of the full complexity of our cells is the main reason why medical drugs often fall short of producing the desired results.

Systems Biology and Regulatory Networks

Regulation in biological systems proceeds on many levels. DNA contains information to make proteins. The types of proteins and the timing of their production are regulated by special DNA-binding proteins. Transcription and translation are regulated by control molecules. All of these processes involve positive and negative feedback loops. Enzymes and molecular machines, as we have seen, are regulated in a variety of ways, from the autoregulation seen in the "parking brake" of kinesin-1 to the complex, sometimes contradictory regulation of molecular machines involving multiple, co-interacting control molecules. On top of this, the cell surface contains numerous specialized receptors, which are controlled by external chemicals. Once a receptor binds to a chemical target, it undergoes a conformational change, which releases or binds a control molecule, setting off a cascade of feedback loops inside the cell, leading to a "macroscopic" response of the entire cell. These signaling pathways are a large part of what biochemists and cell biologists study today.

The first regulatory pathway that was deciphered is the so-called lac-operon in *E. coli* bacteria, for which Jacques Monod and François Jacob received the 1965 Nobel Prize. *E. coli* can live off of a variety of "foods," one of which is lactose (milk sugar). To break down lactose, three enzymes are needed. One of these enzymes is a molecular machine that pumps lactose into the cell. Since it takes resources to make these enzymes, it would not make much sense to produce them if there were no lactose present. In addition, the pump uses precious ATP. But how does DNA know if lactose is present? If lactose is present in the "broth" surrounding the bacteria, a lactose receptor on the cell's surface becomes activated (it binds to lactose and sets off a chemical signal inside the cell through allostery). This activates the few lactose pumps present at the cell's surface, and they begin to pump lactose into the cell. However, these few lactose pumps are not enough to take in all the lactose that is floating by. What to do? Make more pumps!

The gene that encodes the enzymes needed for lactose digestion is preceded by a DNA sequence called the operon. The operon is a DNA patch to which a control protein (a *repressor protein*) can bind. When the repressor binds to the operon, the RNA polymerase, which transcribes DNA into RNA, is blocked, and transcription cannot proceed. No lactose-digesting enzymes are manufactured.

Lactose, however, can bind to the repressor protein, causing it to change shape (via an allosteric interaction) so that the protein can no longer bind to the DNA operon. Now the RNA polymerase is free to transcribe the lactose genes, and lactose-digesting enzymes and lactose pumps are produced in large numbers. Thus, the interaction between lactose, repressor, and the DNA operon makes sure that lactose-digesting enzymes are only produced when lactose is present. This is how the "computer logic" of our cells works. The manufacture of just about every protein is regulated by similar feedback loops.

The study of how these feedback loops work is called *systems biology*. Where molecular biology takes chemical bonding for granted, systems biology takes molecular biology for granted and treats protein and DNA sequences as interacting mathematical entities—players in the computer program of our cells. In this way, scientists work their way up from atoms to molecules to proteins to networks to systems, and finally to an entire cell. Around the world, there are a number of groups trying to develop *virtual cells*—complete simulations of the regulatory networks of simple cells—to understand in detail how they operate.

An important finding from such studies is that as the complexity of simulated networks increases, surprising new properties emerge. In a 1999 paper in the journal *Science*, Upinder S. Bhalla and Ravi Iyengar, two researchers from the Mount Sinai School of Medicine in New York City, simulated interacting signaling networks from their experimental studies of a variety of such networks operating in cells. Bhalla and Iyengar found that by linking different networks together (for example, one network produces a control molecule, which controls an enzyme in another network), new properties emerge that were not part of the individual networks. One such property is *persistent activation*. This property is the persistent production of a protein or control molecule, even after the initial stimulus that caused the production in the first place is long gone. Why would persistent activation be useful? Some processes in our cells take a long time—but at the same time, they may be triggered by a short-lived stimulus. Examples include development, where stem cells need to transform into blood, kidney, or brain cells. Another process where persistent activation is important is the formation of memory. To remember something, our brains must make physical changes to the structure and interaction between brain cells. These changes are triggered by sensory impressions, which become translated into chemical signals. Sensory impressions do not last forever, and neither do the chemical signals derived from them. Yet, we need to remember. Persistent activation makes this possible.

Persistent activation requires a switch, which once flipped, stays on for a long time. At the same time, the switch should not react too easily. After all, creating a permanent memory or turning a stem cell into a brain has serious consequences. One would not want these things to happen unless they were absolutely necessary. Therefore, the switch needs to flip on only if the signal is strong enough and persistent enough. Could simple molecular switches do the job? Molecular switches do not typically send sustained signals. Once they are switched on, they send their signal (releasing a molecule, for example), and that's it.

A better strategy is to have a self-activating feedback loop. This is what Bhalla and Iyengar found. When signaling networks interact in certain ways, they can create a situation where a chemical signal can become locked into a high-activity state—or in other words, where a signaling molecule is continuously produced in high numbers for a long time. This will only happen in response to strong-enough chemical signals. Once the network is activated, the activation is persistent, that is, it will stay activated until another signal switches it off. This *bistability* (in which the network has two stable states: low and high), the duration of the on state, and the threshold concentration needed to activate the network are emergent properties of the interacting networks. Like linking electronic transistors into a larger network in a computer, linking molecular switches together in an interacting network can create more complex functionality.

The Physicist and the Biologist

We have broken down life to its smallest parts: DNA, proteins, enzymes, molecular machines. The idea that we can understand how something works by reducing it to its parts is natural for a physicist. Some consider physics to ultimately be the quest to break things down to smaller and smaller constituents, until we find the one constituent or equation that explains everything. Biologists know that this approach does not work when studying life. There is no "life atom" or one formula to explain life.

As we have seen, with the help of statistical mechanics and nanoscience, we can decipher how the directed activity of molecular machines *emerges* from the underlying atomic chaos. But understanding these machines is still a long way from providing a full understanding of how a living cell works. The next step is to understand how these machines interact in complex signaling and regulation networks. Moving to this level of understanding brings with it its own emergent properties, moving us closer to understanding how life works.

Understanding the parts is crucial, but parts by themselves are not always sufficient to explain the whole. Complex interactions between parts create new processes, structures and principles that, while based materially on the underlying parts, are conceptually independent of them. This insight is what we call holism.

For reasons mysterious to me, there exists a great debate between scientists of various stripes of what the correct approach to science should be—reductionism or holism? As I hope this book has shown, reductionism is essential if we want to understand life. Without it, scientists would have long ago stopped looking at smaller and smaller scales and would have missed the marvels of molecular machinery. At the same time, molecular machines don't explain everything. Scientists must still answer the questions of how these machines interact and what roles they play in the complexity of the cell. The ultimate goal is always to explain the totality of life's processes, from molecules to cells to organisms. Having taken the toy apart, we want to put it back together again. This is the way we learn how things work. Thus reductionism and holism are two sides of the same coin—they are both parts of what good science ought to be.

The battle between holism and reductionism is, in some sense, the modern extension of the ancient battle of the atomists and vitalists. Postulating the existence of perpetually moving atoms, the ancient atomists explained the activity and change they saw in the world as the ultimately impenetrable interactions between atoms. The vitalists adopted a top-down view and declared that it is impossible to reduce life to physical forces, because seen from the outside, life appeared so different and so mysterious compared with the inanimate world. The fault lines of reductionism and holism run through all of science, but especially biology. Ecologists, for example, must think holistically about the interactions of many organisms in a complex environment, while at the other end of the spectrum, molecular biologists and biophysicists look at the smallest possible units of life.

Ernst Mayr was a staunch defender of biological holism and one of the great evolutionary biologists of the twentieth century. Among his many achievements, he provided the best modern definition of species, the idea that species are separated by the inability to interbreed. Mayr wrote a number of wonderful books on the history of biology and evolution and was one of the most spirited defenders of Darwin's theory. But he had (in my opinion) one curious flaw: He hated physicists.

More generally, Mayr deeply disliked any reductionist approaches to biology. In a 2004 paper, he went as far as to make this claim: "To the best of my knowledge, none of the great discoveries made by physics in the twentieth century has contributed anything to an understanding of the living world." Considering the advances we have discussed in this book, which involve twentieth-century physics such as fluorescence spectroscopy, nanotechnology, and X-ray diffraction, this is a curiously uninformed statement by such a great scientist. It seems that he was genuinely concerned that his beloved biology might be taken over by physicists. This fear was and remains unfounded.

In all fairness, Mayr was able to explain more clearly than anyone else why there are differences between biology and physics. For him these differences were in the degree of complexity, the role of chance, the importance of evolutionary history, and the treatment of species as populations. As we have seen, chance and complexity are basic attributes of life. He was quite correct, for example, that much of biology is a question of contingency, of frozen accidents. Biophysics, for example, may explain how a ribosome translates the genetic code into a protein product, but the actual genetic code seems to be a pure accident. There seems to be no reducible physical reason why the genetic letters UUG (corresponding to the RNA molecular bases uracil-uracil-guanine), should translate into a leucine protein subunit, while UGU should translate into cysteine. In most of physics, we don't have such frozen historical accidents. Let copper crystallize from a melt, and it will always crystallize into a face-centered cubic crystal structure. The energy levels in every hydrogen atom are identical. The superconducting transition temperature of mercury is always the same. These things can be predicted from doing quantum mechanics. They happen in accord with fixed laws.

But then again, there are many nonbiological frozen accidents in our universe. Our sun, the earth, the moon, and every mountain on our planet are the result of the vagaries of history. But none of them lies outside a physical explanation. We know the mechanisms that form stars, planets, and mountains. We just cannot predict that a particular mountain will be in a particular place a billion years from now.

For most scientists, philosophical debates over holism versus reductionism are a nonissue. Even most physicists understand that we need to put the parts back together again. Many physical properties, such as elasticity, conductivity, or transparency, arise "holistically" from the interactions of large numbers of atoms. Statistical mechanics, as we saw in Chapter 3, was born to explain the emergence of holistic laws from the reductionist picture of swirling atoms.

Science works on many levels. For a living organism, we may start at quarks and electrons. Using these, we can, in principle, predict the properties of nuclei and atoms. Once we have atoms, we can, in principle and with difficulty, explain the properties of molecules. But even at this point, the connection between the level of quarks and that of molecules is weak at best. We *can* understand many things about molecules by determining their atomic structure, but the quark structure is already too far removed to yield much insight or even a useful explanation for the properties of a molecule. As we move further along, these links become ever more tenuous, until there is really no meaningful conceptual connection between a highly complex entity and the most fundamental levels of matter and energy.

The difficulty in understanding biology is that it operates on many of these levels: from molecules to ecosystems. All of these levels contribute to the understanding of what life (or rather, living) is, and they are all important. As a physicist, I am most fascinated by the levels that connect life to physics, but I am aware that this is just a small part of the complexity of life.

Cows and Quarks

At this point, you might object to my assertion of no meaningful conceptual connection between a complex entity and the most fundamental levels of matter and energy. How did I arrive at this conclusion?

I recently had a discussion on holism versus reductionism with my colleague and friend, Sean Gavin, a theoretical nuclear physicist at Wayne State University. Sean told me that he had listened to a talk by Steven Weinberg, a strong supporter of research in particle physics. As such, Weinberg made his usual argument that particle physics is fundamental to all other sciences, even chemistry and biology. As a physicist, I understand that there is nothing more fundamental than to find out what matter is made of and what forces determine the shape of our universe. This is important work. If we want to learn what our universe is all about—if we want science to progress in the long run—we need to join Weinberg and support the work of the particle physicists. With the recent start-up of the Large Hadron Collider in Geneva, Switzerland, the largest particle accelerator currently in existence, we can expect many new and surprising findings about the deep fundamental structure of our universe. Sean and I, and just about every physicist I know, would agree on this point. But then, Sean said something very funny and to the point: "But how do you predict a cow from particle physics?" A great question!

Is it just too complicated to predict the existence of cows from particle physics, or is it fundamentally impossible to predict a cow from the properties of quarks and electrons? A cow is made of molecules (many of which we discussed in this book). These molecules are made of atoms. The properties of these molecules can, with some difficulty, be reduced to the properties of the atoms they are made of. Atoms are made of quarks, which are held together by gluons (quarks plus gluons form protons and neutrons in the atom's nucleus), and electrons. The chemical properties of different atoms are due to the arrangement of electrons around the nucleus. Thus, we could say that a cow can be explained by particle physics, since quarks and electrons (and the forces acting between them) make atoms with different properties, which in turn make molecules, which in turn make cells, which in turn make cows.

Somewhere along the line, however, we lost sight of why there is such a thing as a cow. To say a cow is explained by what it is made of is to say that bricks explain a house. A better answer is that a cow is the result of evolution—a process made possible by the underlying material reality of particles—but which is essentially unpredictable. If we were to rerun the tape of life, would a cow reemerge? Nobody knows—it is likely that something *like* a cow could emerge again, but the new being might have six legs and only two stomachs. Thus there is no formula for "cow" based on the laws of particle physics. Particle physics may be necessary to make a cow (because we need atoms and molecules), but it is clearly not sufficient.

However, all material objects in this universe are based on the particle physics we know. But if a different universe were to exist, the laws of particles might be quite different from our laws. As long as these laws permit the creation of complicated structures, they may lead to the emergence of something we could justifiably call a cow. The concept "cow" is completely independent of the particular properties of quarks and electrons. A philosopher would say that a cow is not explained by particles, because particles cannot give a reason for the cow's "cow-ness."

What about the reality of molecular machines? As was the case with quarks and electrons, we need to understand molecular machines to understand life on this planet. But we have seen that the molecular machinery of most organisms is not unique to a particular animal or plant. Thus, we cannot derive a cow or any other animal from molecular machines, either. Does this make our insight into molecular machines useless? Do molecular machines tell us nothing about whole organisms? No, molecular machines tell us more than just how cells work. By their similarity in all life on earth, they tell us of evolution and life's unity; by their ability to tame chaos, they tell us a creative universe is only possible through chance and necessity; by their ability to be regulated and to regulate, they tell us that life is matter and program; and by their incessant activity, animated by the molecular storm, they tell us that life is a process, not a thing.

I wager these things would not be so different in a different universe. But then again, who can prove me wrong?

Epilogue

Life, the Universe, and Everything

WE HAVE COME A LONG WAY—FROM THE VITAL FORCES of the ancients to the molecules of molecular biologists and biophysicists. If we were seeking the "life force," the force that animates life, then our search has been successful. This animating force is the random force of atoms, the jittering afterglow of the creation of the universe. The molecular machines, which take this undirected force and give it direction, embody the tight embrace of chance and necessity and are themselves the product of this embrace. Sculpted by evolution, the molecular machines of our bodies tame the molecular storm and turn it into the dance of life.

The universe is the child of chance and necessity. Every star and galaxy, planet and mountain, microbe and elephant is a testament to the interaction between these two basic tendencies of nature. Should this view of the universe, as informed by modern science, influence how we think about ourselves? On one hand, maybe not. Life happens on many levels, from colliding atoms to the mind of a genius. The molecular machines are part of who we are, but they do not determine who we are. We are intelligent, creative beings, a natural extension of the creativity of the universe, but we are not determined by nature. While based on machines, we are not machines ourselves.

On the other hand, science has allowed us to learn something very profound about ourselves. Life is a wonderful molecular mechanism. This should make us admire life even at its most "primitive." Even a virus is a miracle of nature. Humans are part of that same nature—and moreover, we are the most miraculous part of it. We are all the same, and at the same time, we are very different. By *necessity*, we are all bound to the unity of life, but by *chance*, we are all unique. We are supposed to be here—in one shape or another.

If there is life elsewhere in the universe, it will be based on molecular machines. As far as we can determine, the laws of physics are the same everywhere. Everywhere, the nanoscale is the special scale where energies can be easily transformed, providing the potential for autonomous nanoscale machines. Even the humblest living organism is incredibly complex. To attain such complexity, the organism must consist of many interacting parts. These parts must be small, active, varied and complex—only molecules can fit the bill.

To understand the world as a whole, we need to abandon our linear, deterministic thinking. The complexity of life, of our minds, of human society, is the result of adding a dash of randomness to the rules of the game—a game that is played on a network of complex relationships, a game full of emergent properties. I believe (but I cannot prove) that life was inevitable in our large and ancient universe. Consider that life on Earth contains only a tiny fraction of all the matter in the universe. Even if there were millions of inhabited worlds in every galaxy in the universe, the total amount of matter contained in all living beings would still only be a miniscule fraction of all the matter in the universe.

The universe is not a victim of the second law of thermodynamics. If this were so, the universe would just contain diffuse nebulae of hydrogen and helium. But this is not the case. Before life appeared, gravity acted to concentrate atoms. Stars cooked up heavier elements. Planets provided a surface where atoms could be concentrated further, which enabled the creation of complex molecules. The universe is 13.75 billion years old. It had plenty of time and plenty of matter to come up with life *somewhere*. Considering the inherent drive of matter to form ever more complex structures, life seems inevitable.

Of course, there are many who refuse these findings of modern science. They would like to maintain a view of themselves that puts them apart from nature and apart from the universe. In a memorable passage from his 1925 essay "What I Believe," the philosopher Bertrand Russell contrasted such a view with the vision that science has provided: "Even if the open windows of science at first make us shiver after the cozy indoor warmth of traditional humanizing myths, in the end the fresh air brings vigor, and the great spaces have a splendor all their own." To which I would add that once we learn more about science, we find that this shiver is the shiver of excitement—excitement over the grandeur of our universe and our astounding ability to understand a small, but growing corner of it.

Many people express incredulity that something like a human could be the result of the "blind forces" of chance and necessity. They want to believe that the creation of complex structures requires the planning mind of a designer. But how does a mind invent? How do new ideas arise? Are new ideas not chance events, popping into our heads like uninvited houseguests? Could not the same molecular storm that animates our cells sometimes shake our thoughts and create sudden insights? Such random thoughts may cause us to create new connections between seemingly unconnected experiences, leading us to think outside the box. This model of human thought makes sense to me. How could new ideas come about any other way? Where would they come from? From the outside? No, we know that ideas are generated by brains, which are complex networks of biological cells, communicating via chemical and electrical signals. The only way to generate new ideas is by involving some degree of randomness. Even if we invoke an all-powerful mind to explain the origin of the universe or of life, we are thrown back to the same basic forces of chance and necessity. Even this all-powerful mind would have to depend on them.

Are we getting closer to understanding all there is to understand? One hundred and fifty years ago, Charles Darwin threw up his hands and exclaimed, "I feel most deeply that this whole question of Creation is too profound for human intellect. A dog might as well speculate on the mind of Newton! Let each man hope and believe what he can." I understand the sentiment. We are *still* far away from penetrating the mystery of mysteries. But we have come much, much closer.

If we do not yet completely understand life, it is because life is incredibly complex. How does one combine the fundamental parts of life—DNA, enzymes, molecular machines—to create a Shakespeare or an Einstein?

Faced with such mysteries, many people want to throw up their hands like Darwin and declare that it is not possible to explain life after all. Yet, as we have seen throughout this book, science can turn darkness into light and can reveal deep secrets of life. We have seen that life is governed by chance and necessity.

Glossary

actin A long, fibrous protein that is part of the "skeleton" of a cell. Also acts as track for myosin and forms the fibers on which myosin II pulls in muscles.

activation barrier Energy barrier that molecules have to overcome when they change shape or react with one another.

ADP (adenosine diphosphate) Product of removal of one phosphate group from ATP, an energy-carrying molecule that is used in cells to move chemical energy around. ADP consists of a nucleotide (adenine) with two phosphate groups attached.

allostery The ability of some enzymes to change shape and functionality in response to binding a control molecule. Allostery constitutes the basis of regulation in cells.

amino acid Smallest unit of a protein. Proteins consist of a various combinations of the twenty amino acids that are used in nature.

amphiphilic A molecule that has both hydrophilic and hydrophobic characteristics.

animism The belief that everything has a soul and is alive.

atomic force microscopy (AFM) Type of scanning probe microscopy, which measures small forces between a sharp tip and a surface. AFM can provide high-resolution images or can be used to measure forces between molecules.

atomism The belief that everything is made of small, indivisible, and perpetually moving particles.

atoms Smallest chemical units, composed of a nucleus (made of protons and neutrons) and a cloud of electrons. Electrons are responsible for chemical bonding.

ATP (adenosine triphosphate) Energy-carrying molecule, used in cells to move chemical energy around. Consists of a nucleotide (adenine) with three phosphate groups attached.

ATP hydrolysis Reaction of ATP with water, which detaches one of its phosphate groups and liberates a large amount of energy. End product: ADP (adenosine diphosphate).

ATP synthase Sophisticated rotary molecular machine located in mitochondria. Uses a proton gradient to recharge ATP.

ATPase Enzyme that breaks down ATP. Part of almost all molecular machines.

bit Minimum quantity of information; information contained in a "yes" or "no" answer (or "1" and "0").

Brownian motion The random motion of small particles as the result of many collisions with molecules in the air or in a liquid.

Brownian ratchet A molecular machine that moves in a specific direction via a directed diffusion process. Does not violate the second law of thermodynamics because in order to work, energy is supplied to the ratchet to periodically detach the machine from the track on which it moves. This energy is then dissipated.

chance and necessity The basic principles responsible for everything there is. Chance arises from quantum mechanics and the molecular storm, while necessity is due to physical laws.

chaperonin Proteins that help other proteins fold into the correct shape.

chromosome A bundle of DNA in the cell nucleus.

codon A "word" in the genetic code; consists of three nucleotide letters. Each codon encodes one amino acid according to the genetic code.

collagen Fibrous proteins. Part of the extracellular matrix, giving structure to animal bodies.

complexity Attribute of a system that is composed of many interacting parts and that exhibits emergent properties.

cooperativity The property of some processes wherein several parts must act together simultaneously for the process to occur.

diffusion Random motion of molecules or atoms. On average, diffusion leads to the movement of particles from a region of high concentration to a region of low concentration.

dissipation Degradation of usable (free) energy into unusable energy (heat).

DNA (deoxyribonucleic acid) A long, double-helical molecule located in the cell nucleus; stores the sequences to make proteins and directs the development of the cell.

domain Part of a large molecule.

dynein Family of molecular motors that move on microtubules.

emergence Arising of properties resulting from the interaction of many parts; the emergent properties are not properties of the parts by themselves.

energy A propensity to perform work. The unit of energy is the joule (J).

energy landscape Conceptual idea of a multidimensional landscape representing how the energy of a protein changes as it changes shape. Each location in the landscape corresponds to a specific protein shape, while the height of the landscape represents the energy associated with the shape.

energy transformation The change of energy from one type into another, for example, from chemical to kinetic energy, as in a car.

entropic forces Forces that are not due to the reduction in energy, but are due to an increase in entropy. Examples are depletion forces and hydrophobic forces.

entropy The degree to which energy is dispersed. Often equated with "disorder."

enzyme Protein-based catalyst; protein that can facilitate a reaction by lowering the reaction's activation barrier.

equilibrium In thermodynamics, the state at which entropy is maximum, free energy is minimum, and no more useful work can be extracted from a system. Characterized by uniform temperature and pressure.

evolution Process by which new proteins, molecular machines, or organisms arise. Novelty is produced by mutations, which are then acted upon by selection. Acts like a "ratchet," leading to increasing complexity.

feedback loop The interaction of products of a process with the precursors of a process via allostery, whereby the amount of product is controlled. Feedback loops are the basic component of the chemical "circuitry" that regulates cells.

Feynman's ratchet An automated version of Maxwell's demon, which could extract energy from a system at equilibrium, thus violating the second law of thermodynamics. Shown to be impossible.

first law of thermodynamics Law of energy conservation, which explicitly includes heat as a type of energy.

fluorescence Emission of longer-wavelength light by a molecule in response to excitation by shorter-wavelength light.

fluorophore Molecule that emits visible light when excited by light of shorter wavelength (fluorescence). Used to label molecules so that their behavior can be observed in an optical microscope.

food calorie Energy unit, equals 4,184 joules.

force A push or pull that either deforms an object or changes its state of motion. The unit of force is the newton (N).

free energy Energy that is available to do work. Equals total energy minus unusable, dispersed energy. The latter is given by the product of entropy and temperature.

Gedankenexperiment Posing a hypothetical situation to see if a physical theory makes sense. From a German word meaning "thought experiment."

gene A unit of hereditary information. A complete sequence of information in the genome, encoded in DNA.

genetic code The code that DNA uses to encode amino acid sequences of proteins.

head Term for the motor domain of a motor protein, specifically, the part on which it walks.

helicase Molecular machine that untwists and separates DNA strands during replication.

holism The view that systems have to be studied as a whole, without being broken into parts.

hydrophilic Molecules that lower their free energy when dissolved in water. Can either form hydrogen bonds with water, or are charged.

hydrophobic Molecules that experience an increase in free energy when placed in water, and do therefore not dissolve in water.

irritability The property of living tissue to react to external stimuli such as electrical currents.

kinesin Family of molecular motors that move on microtubules. Responsible for moving cargo or assisting in cell division.

kinetic theory A precursor to statistical mechanics; the application of statistics to the motion of atoms in gases.

Landauer's limit The minimum energy required to erase one bit of information.

laser tweezers Laser-based force measurement technique using a small bead suspended in focused laser light.

ligase Enzyme that splices parts of DNA together.

lipids Amphiphilic molecules, which form micelles or vesicles and are the main constituent of cell membranes. Consist of a hydrophobic fatty acid bound to a hydrophilic head group.

loose coupling The ability of a molecular motor to move variable distance for each ATP consumed. Such a motor must periodically detach from the track and move as a Brownian ratchet.

Loschmidt's demon A hypothetical creature that could reverse time.

macrostate The externally measurable state of a system, which is necessarily an average over many atoms and molecules. Usually described in terms of macroscopic quantities like temperature or pressure.

Maxwell's demon Hypothetical creature that could sort molecules and violate the second law of thermodynamics. Shown to be impossible.

messenger RNA (mRNA) RNA that carries the information to make a protein from DNA in the cell nucleus to the ribosomes outside the nucleus.

micelle Spherical assembly of amphiphilic molecules.

micrometer One-millionth of a meter. Typical size of a bacterium.

microstate The exact state of a system; includes the location and speeds of all atoms in the system.

microtubules Stiff, protein-based fibers that make up part of the skeleton of the cell. Kinesin and dynein move on microtubules. Microtubules play an important role during cell division.

mitochondria Power plants of cells. This is where sugar is broken down and the energy contained is transferred to ATP.

mitosis Cell division.

MMP (matrix metalloprotease) An enzyme that breaks down the extracellular matrix and frees up cells so they can move. Some MMPs have been shown to be molecular machines that derive their energy from eating collagen.

molecular machine Molecule that can transform energy from one form to another.

molecular motor Molecular machine that transforms chemical into mechanical energy.

molecular storm Random thermal motion of atoms and molecules.

molecule Tightly bonded assembly of at least two atoms. In biological cells, some molecules, such as DNA or proteins, can be very large, containing many thousands of atoms.

mutation Irreversible change in a gene; the result of a chemical reaction of DNA with an ion or a radical (atoms, molecules, or ions with unpaired electrons).

myosin Family of molecular motors that move on actin. Myosin II is responsible for muscle contraction, while myosin V transports cargo within cells.

nanobots Hypothetical nanoscale machines that can perform complex tasks, especially with medical applications.

nanometer One-billionth of a meter. Typical size of a small molecule.

nanoscale The scale of the nanometer.

nanoscience The science of nanoscale objects.

nanotechnology Technology that involves nanoscale structures or materials.

normal distribution Distribution expected for measurements that depend on several statistically independent influences. This universal distribution is found in physics, biology, economics, and other fields.

nucleotide Molecule that makes up a "letter" in the genetic code. Part of DNA and RNA. There are four nucleotides in DNA: adenine, guanine, thymine, and cytosine, denoted by the letters A, G, T, and C.

open system Systems that can exchange energy or matter, or both, with the environment; as opposed to isolated systems.

perpetuum mobile Machine that could violate the first, or at least the second, law of thermodynamics. Found to be impossible.

polymerase Molecular machine that makes (polymerizes) DNA or RNA.

power Rate of transformation of energy into a different form, or rate of work performed. Energy per second. The unit of power is the watt (W); 1 watt = 1 joule per second.

pressure Force per unit area. In a gas, pressure is the result of many collisions of atoms with a wall.

probability The likelihood of the occurrence of an event; given by the number of ways the event could occur, divided by all possible outcomes.

processivity The ability of a molecular motor to move on a molecular track for long distances without detaching.

protein A large molecule consisting of a folded strand of amino acids connected by peptide bonds. Proteins fold into specific shapes, which allow them to fulfill many tasks in cells, including acting as enzymes, molecular machines, or structural elements (collagen, actin, etc.).

protein folding Physical process in which a chain of amino acids folds into its lowest free-energy state to form a functional protein.

quantum mechanics Physical theory that describes atoms, electrons, and subatomic particles. Fundamentally based on probabilities.

reductionism A method in which systems are analyzed by breaking them into smaller parts.

replication The copying of DNA during cell division.

ribosome RNA-based machine in our cells that produces proteins according to instruction contained in DNA.

RNA (ribonucleic acid) Molecule that can both carry information and act as a catalyst. Plays many roles in cells, from information carrier to protein production machines (ribosome).

scanning probe microscopy Type of microscopy where a sharp probe "feels" a surface to generate an image or to measure surface properties.

scanning tunneling microscopy A type of scanning probe microscopy that measures tunneling currents between a sharp metallic needle and a conducting surface. Can provide images of atoms.

second law of thermodynamics Law that states that in a closed system, entropy can never decrease, but generally tends to increase during energy transformations. There are many alternative formulations of this law.

statistical mechanics The application of statistical methods to the motions of atoms and molecules. Averaging over many atoms and molecules leads to the macroscopic science of thermodynamics.

systems biology Mathematical theory of regulation and computation in living cells.

temperature The average kinetic energy of atoms or molecules in a system at equilibrium.

thermodynamics The science of heat and energy, volume, pressure, and temperature.

tight coupling The exact coupling of ATP hydrolysis to the motion of a molecular motor. Typically implies that one ATP is hydrolyzed for each step taken by the motor. Associated with motors that have at least one part of the motor attached to the track at all times. As opposed to loose coupling.

Topoisomerase Enzyme that cuts and reattaches DNA strands to relieve strain and keep DNA from tangling during replication.

transcription The transferring of information from DNA to messenger RNA.

transfer RNA (tRNA) RNA–amino acid complexes that contain the key to translate the genetic information into a protein.

transition state "Uncomfortable" state that molecules find themselves in temporarily as they transform from one stable form to another. Energy associated with the transition state is the activation barrier.

translation Translating genetic information into a protein. Happens in the ribosome, with the aid of transfer RNA (tRNA).

vesicle Spherical assembly of lipids; consists of a double-walled sphere separating a water-filled chamber from water on the outside of the vesicle.

vitalism The belief that life is associated with special forces.

work The product of force and distance; the expenditure of energy when force acts over a certain distance to move an object. The unit of work is the joule (J), which is equal to a newton-meter (N·m).

X-rays Highly energetic electromagnetic radiation, generated by bombarding high-speed electrons into a metal target.

Sources

This a partial list of sources I used to write the book. Many of these sources were used in multiple chapters. However, to save space, each source is mentioned only once for the chapter in which it was used first. The references are listed alphabetically by last name of author. I listed only the sources that are easily accessible. Unfortunately, many technical papers are both inaccessible to the public and often incomprehensible to the nonexpert.

INTRODUCTION

Monod, Jacques. *Chance and Necessity*. Translated by Austryn Wainhouse. New York: Alfred A. Knopf, 1971.

Schrödinger, Erwin. *What Is Life?* Canto ed. Cambridge: Cambridge University Press, 1992.

Thomson, D'Arcy. *On Growth and Form*. Canto ed. Cambridge: Cambridge University Press, 1992.

CHAPTER 1

Alioto, Anthony. *A History of Western Science*. Englewood Cliffs, N.J.: Prentice-Hall, 1986.

Aristotle. *De Anima*. In *The Basic Works of Aristotle*. New York: Modern Library, 2001.

———. *Physics*. In *The Basic Works of Aristotle*. New York: Modern Library, 2001.

Cahan, David. *Hermann von Helmholtz and the Foundations of Nineteenth-Century Science*. California Studies in the History of Science. Berkeley: University of California Press, 1994.

Darwin, Charles. *The Origin of Species*. Available at Literature.org, the Online Literature Library, www.literature.org/authors/darwin-charles/the-origin-of-species/.

Debus, Allen George. *Man and Nature in the Renaissance*. Cambridge Studies in the History of Science. Cambridge: Cambridge University Press, 1978.

Descartes, René. *Discourse on Method*. Available at Literature.org, the Online Literature Library, www.literature.org/authors/descartes-rene/reason-discourse/.

Gay, Peter. *The Enlightenment: The Rise of Modern Paganism*. New York: W. W. Norton, 1995.

Granger, Herbert. *Aristotle's Idea of the Soul*. Boston: Kluwer, 1996.

Hankins, Thomas. *Science and the Enlightenment*. Cambridge Studies in the History of Science. Cambridge: Cambridge University Press, 1985.

Harris, Henry. *The Birth of the Cell*. New Haven, Conn.: Yale University Press, 2000.

Helmholtz, Hermann. *Über die Erhaltung der Kraft*. Available at The Internet Archive, www.archive.org/details/berdieerhaltung01helmgoog.

Henry, John. *The Scientific Revolution and the Origins of Modern Science*. Studies in European History. New York: Palgrave Macmillan, 2008.

Holmes, Richard. *The Age of Wonder: How the Romantic Generation Discovered the Beauty and Terror of Science*. New York: Pantheon, 2009.

Hooke, Robert. *Micrographia*. Available at Project Gutenberg, www.gutenberg.org/files/15491/15491-h/15491-h.htm.

La Mettrie, Julien. *Machine Man and Other Writings*. Edited by Ann Thomson. Cambridge: Cambridge University Press, 1996.

La Mettrie, Julien, and Aram Vartanian. *La Mettrie's L'homme Machine: A Study in the Origins of an Idea*. Princeton, N.J.: Princeton University Press, 1960.

Lenoir, Timothy. *The Strategy of Life: Teleology and Mechanics in Nineteenth-Century German Biology*. Chicago: University of Chicago Press, 1989.

Lindberg, David. *The Beginnings of Western Science: The European Scientific Tradition in Philosophical, Religious, and Institutional Context, Prehistory to A.D. 1450*. Chicago: University of Chicago Press, 2008.

Lucretius. *De Rerum Natura*. Available at The Internet Classic Archive, http://classics.mit.edu/Carus/nature_things.html.

McKendrick, John Gray. *Hermann Ludwig Ferdinand von Helmholtz (Masters of Medicine)*. New York: Longmans, Green & Co., 1899.

Mendelsohn, Everett. *Heat and Life: The Development of the Theory of Animal Heat*. Cambridge, Mass.: Harvard University Press, 1964.

Newton, Isaac. *Opticks*. Available at Project Gutenberg, www.gutenberg.org/files/33504/33504-h/33504-h.htm.

Rubenstein, Richard. *Aristotle's Children: How Christians, Muslims, and Jews Rediscovered Ancient Wisdom and Illuminated the Middle Ages*. New York: Houghton Mifflin, Mariner Books, 2004.

Russell, Bertrand. *A History of Western Philosophy*. New York: Simon & Schuster, 1972.

Shelley, Mary Wollstonecraft. *Frankenstein—or the Modern Prometheus*. Available at Literature.org, the Online Literature Library, www.literature.org/authors/shelley-mary/frankenstein/.

CHAPTER 2

Chardin, Teilhard. *The Phenomenon of Man*. Available at The Internet Archive, www.archive.org/details/phenomenon-of-man-pierre-teilhard-de-chardin.pdf.

Eigen, Manfred, and Ruthild Winkler. *Das Spiel—Naturgesetze steuern den Zufall*. Serie Piper, 1996; English edition: *Laws of the Game: How the Principles of Nature Govern Chance*. Princeton, N.J.: Princeton University Press, 1993.

Judson, Horace Freeland. *The Eighth Day of Creation: Makers of the Revolution in Biology*. Plainview, N.Y.: Cold Spring Harbor Laboratory Press, 1995.

Kline, Morris. *Mathematics for the Nonmathematician*. New York: Dover, 1985.

———. *Mathematics in Western Culture*. New York: Oxford University Press, 1964.

Mainzer, Klaus. *Der kreative Zufall—Wie das Neue in die Welt kommt*. Munich, Germany: C. H. Beck, 2007.

Mlodinow, Leonard. *The Drunkard's Walk: How Randomness Rules Our Lives*, New York: Pantheon, 2008.

Moore, Walter. *Schrödinger: Life and Thought*. Cambridge: Cambridge University Press, 1992.

Timoféeff-Ressovsky, N. W.; K. G. Zimmer; and M. Delbrück. *Über die Natur der Genmutation und der Genstruktur* (the "green pamphlet"). In *Nachrichten von der Gesellschaft der Wissenschaften zu Göttingen*. Berlin: Weidmannsche Buchhandlung, 1935. Available at www.ini.uzh.ch/~tobi/fun/max/timofeeffZimmerDelbruck1935.pdf.

Wykes, Alan. *Doctor Cardano, Physician Extraordinary*. London: Muller, 1969.

CHAPTER 3

Feynman, Richard. *The Feynman Lectures of Physics*. Boston: Addison Wesley Longman, 1970.

Lindley, David. *Boltzmann's Atom: The Great Debate That Launched a Revolution in Physics*. New York: Free Press, 2001.

CHAPTER 4

Drexler, K. Eric. *Engines of Creation 2.0*. E-book. WOWIO, 2006. Available at www.wowio.com/users/product.asp?BookId=503.

Feynman, Richard. "There's Plenty of Room at the Bottom." *Caltech Engineering and Science* 23 (February 1960): 22–36. Available at www.zyvex.com/nanotech/feynman.html.

Nelson, Philip. *Biological Physics: Energy, Information, Life*. New York: W. H. Freeman, 2008.

Philips, Rob, and Stephen R. Quake. "The Biological Frontier of Physics." *Physics Today* (May 2006): 38–43.

CHAPTER 5

Feynman, Richard. *Feynman Lectures on Computation*. Reading, Mass.: Perseus, 2000.

Gamow, George. *Mr. Tompkins in Paperback*. Canto ed. Cambridge: Cambridge University Press, 1993.

Mahon, Basil. *The Man Who Changed Everything: The Life of James Clerk Maxwell*. New York: Wiley, 2003.

CHAPTER 6

Astumian, R. Dean. "Making Molecules into Motors." *Scientific American* (July 2001).

Astumian, R. Dean, and Peter Hänggi. "Brownian Motors." *Physics Today* (November 2002): 33–39.

CHAPTER 7

Boyer, Paul. "Energy, Life, and ATP." Nobel Lecture, December 8, 1997. Available at http://nobelprize.org/nobel_prizes/chemistry/laureates/1997/boyer-lecture.html.

Cooper, Geoffrey. *The Cell: A Molecular Approach*. Washington, D.C.: ASM Press, 1997.

Hoagland, Mahlon, and Bert Dodson. *The Way Life Works*. New York: Three Rivers Press, 1995.

Walker, John. "ATP Synthesis by Rotary Catalysis." Nobel Lecture, December 8, 1997. Available at http://nobelprize.org/nobel_prizes/chemistry/laureates/1997/walker-lecture.html.

CHAPTER 8

Paley, William. *Natural Theology*. Available at "The Complete Works of Charles Darwin Online," http://darwin-online.org.uk/content/frameset?itemID=A142&viewtype=text&pageseq=1.

CHAPTER 9

Alon, Uri. *An Introduction to Systems Biology*. Boca Raton, Fla.: Chapman & Hall/CRC, 2007.

Margulis, Lynn, and Dorion Sagan. *What Is Life?* Berkeley: University of California Press, 1995.

Mayr, Ernst. *This Is Biology: The Science of the Living World*. Cambridge, Mass.: Harvard University Press, 1997.

Suggested Reading

This is an annotated list of books that I highly recommend for further reading. They may help clarify topics mentioned in the book, or continue a topic where the book left off.

ATOMISM

Greenblatt, Stephen. *The Swerve: How the World Became Modern*. New York: W. W. Norton, 2011. Delightful reading about impact of ancient Greek and Roman atomistic ideas on modern science.

COMPLEXITY

Kauffman, Stuart. *At Home in the Universe: The Search for the Laws of Self-Organization and Complexity*. New York: Oxford University Press, 1996. Thought-provoking ideas about self-organization, complexity, and the origin of life, from one of the best-known complexity theorists.

Waldrop, M. Mitchell. *Complexity: The Emerging Science at the Edge of Order and Chaos*. New York: Simon & Schuster, 1992. A highly readable, almost journalistic account of the early days of complexity research. Although the book is older, the topics discussed are just as relevant today as they were in the early 1990s.

DEVELOPMENT

Carroll, Sean. *Endless Forms Most Beautiful: The New Science of Evo Devo*. New York: Norton, 2005. This book and Carroll's *The Making of the Fittest*, below, provide a superb introduction into evolutionary development ("evo devo")—the theory of how bodies get their shapes and how these shapes evolve.

———. *The Making of the Fittest: DNA and the Ultimate Forensic Record of Evolution*. New York: W. W. Norton, 2006. See notes on *Endless Forms Most Beautiful*, above.

EVOLUTION

Dawkins, Richard. *The Ancestor's Tale: A Pilgrimage to the Dawn of Evolution*. New York: Houghton Mifflin, 2004. My favorite Dawkins book (but all his books are recommended). A travel back in time, species by species, to the origin of life.

Pallen, Mark. *The Rough Guide to Evolution*. New York: Penguin Books, 2009. A quick, but surprisingly detailed introduction to evolution. A fun read.

Zimmer, Carl. *Evolution: The Triumph of an Idea*. New York: Harper Collins, 2001. A companion book to the highly recommended PBS TV series. Superb explanations, good writing, and many, many illustrations.

GENETICS

Ridley, Matt. *Genome: The Autobiography of a Species in 23 Chapters*. New York: Harper Perennial, 2000. A great introduction to the human genome. Each chapter covers a chromosome. The writing in this book is excellent.

MOLECULAR BIOLOGY

Harold, Franklin M. *The Way of the Cell: Molecules, Organisms, and the Order of Life*. New York: Oxford University Press, 2003. A very readable, popular introduction to cell biology.

Hoagland, Mahlon, and Bert Dodson. *The Way Life Works*. New York: Three Rivers Press, 1995. This book is also listed in sources, but I list it here again because I believe that everybody who has any interest in biology must have this book. It is a unique combination of humor and cartoons and a serious introduction to molecular biology. This is the best book to get you up to speed.

Lane, Nick. *Power, Sex, Suicide: Mitochondria and the Meaning of Life*. New York: Oxford University Press, 2005. A detailed, but very readable account of how energy is generated in cells, but it also branches out into many fundamental questions, such as why there are two sexes.

Rensberger, Boyce. *Life Itself: Exploring the Realm of the Living Cell*. New York: Oxford University Press, 1996. A well-written, popular introduction to cell and molecular biology.

MOLECULAR MACHINES

Jones, Richard. *Soft Machines: Nanotechnology and Life*. New York: Oxford University Press, 2007. Covers ground similar to *Life's Ratchet*, but with less emphasis on the physics and more emphasis on nanotechnology. A good read.

Nelson, Philip. *Biological Physics: Energy, Information, Life*. New York: Freeman, 2008. Although I already listed this book in my sources, I list it here again because of its importance to *Life's Ratchet*. This book inspired me to write the book in the first place. *Biological Physics* is the most interesting and well-written textbook I have ever read. However, it is quite technical, so buy it only if your calculus and physics are solid.

ORIGIN OF LIFE

Davies, Paul. *The Fifth Miracle: The Search for the Origin and Meaning of Life*. New York: Simon & Schuster, 2000. A very readable introduction to theories about life's origin.

Hazen, Robert. *Genesis: The Scientific Quest for Life's Origin.* Washington, D.C.: Joseph Henry Press, 2007. A very readable, personal account of an origin of life researchers. Up-to-date.

SELF-ORGANIZATION AND PATTERNS IN NATURE

Ball, Philip. *The Self-Made Tapestry: Pattern Formation in Nature.* New York: Oxford University Press, 2001. A detailed and beautifully illustrated account of the spontaneous formation of patterns in nature. A modern update of D'Arcy Thompson's *Growth and Form*.

THERMODYNAMICS AND LIFE

Brown, Guy. *The Energy of Life: The Science of What Makes Our Minds and Bodies Work*. New York: Free Press, 1999. A popular account of how thermodynamics plays into human life, from the thermal motion in our cells to losing weight.

Kurzynski, Michal. *The Thermodynamic Machinery of Life.* The Frontiers Collection. Berlin and New York: Springer, 2006. A very technical, but profound discussion of thermodynamics and life.

Schneider, Eric. *Into the Cool: Energy Flow, Thermodynamics, and Life*. Chicago: University of Chicago Press, 2006. Discusses not only how life and the second law of thermodynamics are compatible, but how the second law is necessary to explain life.

Acknowledgments

A book that starts in ancient Greece and ends up with today's cutting edge research cannot possibly be created by one person alone. A number of people have, directly or indirectly, inspired, helped, worked on, or otherwise enabled me to write this book, and for this I thank them.

First, I thank my lovely and intelligent wife, artist and fellow science enthusiast Patricia Domanski, who spent numerous hours reading, criticizing, editing, and correcting draft after draft, until the last vestiges of meandering sentences, dragging paragraphs, and useless words were removed. If there are any of them left, it is entirely my fault. My discussions with her inspired many ideas in the book, and I must especially thank her for having almost infinite patience with me.

I also thank my parents and teachers, who have instilled a great love for learning in me, a gift that has driven me to always want to learn and experience new things. One result of this gift is this book. I also thank Professor Herbert Granger from the Wayne State University Department of Philosophy, who set me straight on pre-Socratic philosophy; Professor Rafael Fridman (WSU Pathology) and Professor Heinrich Hoerber (now at University of Bristol), who got me started in biology-related research; and my physics colleague, Dr. Takeshi Sakamoto, who answered my many questions about molecular machines.

Last, but not least, I want to thank my agent, Russell Galen, who was so kind to take a chance on me, even though my first attempt at a book proposal was less than stellar, and my editors at Basic Books, T. J. Kelleher and Tisse Takagi, for all their valuable suggestions and support.

Index